SYNCHRONIZATION

A modern introduction to synchronization phenomena, this text presents recent discoveries and the current state of research in the field, from low-dimensional systems to complex networks. The book describes some of the main mechanisms of collective behavior in dynamical systems, including simple coupled systems, chaotic systems, and systems of infinite dimension. After introducing the reader to the basic concepts of nonlinear dynamics, the book explores the main synchronized states of coupled systems and describes the influence of noise and the occurrence of synchronous motion in multistable and spatially extended systems. Finally, the authors discuss the underlying principles of collective dynamics on complex networks, providing an understanding of how networked systems are able to function as a whole in order to process information, perform coordinated tasks, and respond collectively to external perturbations. The demonstrations, numerous illustrations, and application examples will help advanced graduate students and researchers gain an organic and complete understanding of the subject.

STEFANO BOCCALETTI is Senior Researcher at the CNR Institute for Complex Systems. Previously he has been the Scientific Attaché at the Italian Embassy in Israel, Full Researcher at the National Institute of Optics in Italy, and Visiting Scientist or Honorary Professor of seven international universities. He is the editor of four books and Editor in Chief of *Chaos Solitons and Fractals*.

ALEXANDER N. PISARCHIK is Isaac-Peral Chair in Computational Systems Biology at the Center for Biomedical Technology of the Technical University of Madrid. His research interests include chaos theory and applications in optics, electronics, biology and medicine, chaotic cryptography, and communication.

CHARO I. DEL GENIO is Visiting Faculty Member at the University of Warwick and his work primarily focuses on graph theory and complex networks, particularly with an algorithmic or simulation component. Recently he has been applying methods from network research to the study of biological systems, to explain biomolecular mechanisms and design new antimicrobial drugs.

ANDREAS AMANN is a lecturer at University College Cork and his research interests focus on semiconductor physics, lasers, photonics, and, more recently, energy harvesting devices. From a mathematical perspective his work concerns synchronization, time delay, and complex networks.

SYNCHRONIZATION

From Coupled Systems to Complex Networks

STEFANO BOCCALETTI

CNR Institute for Complex Systems, Rome

ALEXANDER N. PISARCHIK

Technical University of Madrid, Spain

CHARO I. DEL GENIO

University of Warwick

ANDREAS AMANN

University College Cork

CAMBRIDGE
UNIVERSITY PRESS

CAMBRIDGE
UNIVERSITY PRESS

University Printing House, Cambridge CB2 8BS, United Kingdom

One Liberty Plaza, 20th Floor, New York, NY 10006, USA

477 Williamstown Road, Port Melbourne, VIC 3207, Australia

314–321, 3rd Floor, Plot 3, Splendor Forum, Jasola District Centre, New Delhi – 110025, India

79 Anson Road, #06–04/06, Singapore 079906

Cambridge University Press is part of the University of Cambridge.

It furthers the University's mission by disseminating knowledge in the pursuit of
education, learning, and research at the highest international levels of excellence.

www.cambridge.org
Information on this title: www.cambridge.org/9781107056268
DOI: 10.1017/9781107297111

© Stefano Boccaletti, Alexander N. Pisarchik, Charo I. del Genio, and Andreas Amann 2018

First published 2018

Printed in the United Kingdom by TJ International Ltd. Padstow Cornwall

A catalogue record for this publication is available from the British Library.

Library of Congress Cataloging-in-Publication Data
Names: Boccaletti, S. (Stefano), author. | Pisarchik, A. N. (Alexander N.),
author. | Del Genio, Charo I., author. | Amann, Andreas, author.
Title: Synchronization : from coupled systems to complex networks / Stefano
Boccaletti (Consiglio Nazionale delle Ricerche (CNR), Rome), Alexander N.
Pisarchik (Centro de Investigaciones en Optica, Leon, Mexico), Charo I.
del Genio (University of Warwick), Andreas Amann (University College Cork).
Description: Cambridge : Cambridge University Press, [2018] | Includes
bibliographical references and index.
Identifiers: LCCN 2017054585 | ISBN 9781107056268 (alk. paper)
Subjects: LCSH: Nonlinear theories. | Synchronization. | Pattern formation
(Physical sciences) | Chaotic synchronization. | Dynamics.
Classification: LCC QC20.7.N6 B63 2018 | DDC 530.15–dc23
LC record available at https://lccn.loc.gov/2017054585

ISBN 978-1-107-05626-8 Hardback

Contents

Preface

Understanding, predicting, and controlling the way complex systems coordinate their dynamics in a cooperative manner have been, and still remain, among the fundamental challenges in modern scientific research. Their overwhelming difficulty stems from two major issues: extracting the proper dynamics of a solitary system, and capturing the complex way through which different systems (or units) interact to function together and in coordination with one another.

This book describes some of the most important mechanisms through which collective behavior of dynamical systems emerges, starting from the case of simple coupled systems, up to chaotic systems, infinite-dimensional systems, space-extended systems, and complex networks.

We focus on synchronization (from the Greek $\sigma\acute{u}\nu$ = together and $\chi\rho\acute{o}\nu o\varsigma$ = time), which literally means "happening at the same time." Synchronization is actually a process where, due to their interactions or to an external driving force, dynamical systems adjust some properties of their trajectories so that they eventually operate in a macroscopically coherent way.

The word *synchronization* appeared first in 1620, at a time when determining longitudes was a challenge for transoceanic voyages. To find a solution to this problem, Christiaan Huygens invented the pendulum clock in 1657. For practical purposes, two clocks were required in general, in case one of the two stopped working properly. So, Huygens studied the behavior of two simultaneously operating maritime clocks, and noticed that they evolved in a synchronized manner and oscillated in the same plane when they were close to one another. Since then, synchronization has been investigated in numerous fields, such as mechanics, chemistry, neuroscience, biology, ecology, and social interactions, to quote just a few examples.

Synchronization is ubiquitous in natural phenomena: the organization of the world, outside and inside us, mostly depends on how different parts, units, and components are able to synchronize. The study of synchronization, as behavior of coupled systems correlated in time, is therefore a fast-developing research topic with applications in almost all areas of science and engineering, ranging from chaotic communication to complex biological and social networks.

After introducing the reader to the basic concepts of nonlinear dynamics, the book explains the main synchronization states, such as complete synchronization, phase synchronization, lag and anticipated synchronization, generalized synchronization, and intermittent synchronization, which happen among coupled systems. Then, we move toward describing the influence of noise on synchronization, and the occurrence of synchronous motion in multistable systems and spatially extended systems, with a description of related effects, including amplitude and oscillation death.

Finally, the book discusses underlying principles of collective dynamics on complex networks, to provide an understanding of how networked systems are able to function as a whole in order to process information, perform coordinated tasks, give rise to parallel computing operations, and respond collectively to external perturbations or noise. We also review recent progress towards establishing a realistic and comprehensive approach capable of explaining and predicting some of the collective activity of real-world networks.

Furthermore, some of the most important applications of synchronization in different areas of science and engineering are given attention throughout this book: electronic circuits, lasers, chaotic communication, and neural networks.

The basic aim is to provide a first approach to synchronization, for readers who are interested in understanding its fundamental concepts and applications in several fields, from students and technicians to scientists and engineers conducting interdisciplinary research, both theoretical and experimental.

1

Introduction and Main Concepts

This chapter is aimed at an audience that is not yet familiar with the area of nonlinear dynamics and its mathematical description. Its goal is to introduce the reader to the terminology and fundamental tenets adopted by the majority of scientists working in this area, as well as to provide theoretical and mathematical tools and background to better understand the subsequent parts of the book.

Concepts of nonlinear dynamics, such as dynamical systems, bifurcations, attractors, and Lyapunov exponents, will be briefly described. However, since the main topic of this book is synchronization, we present these auxiliary topics with the minimum of details and limit ourselves to a rather informal descriptive presentation. A reader interested in further deepening their general knowledge in nonlinear science and its applications is referred to the following books: Schuster and Just 2005; Baker and Gollub 1996; Ott 2002; Strogatz 2015; Fuchs 2013; Guckenheimer and Holmes 1983.

1.1 Dynamical Systems

From atoms to galaxies, at every length scale of study, one can distinguish relatively isolated self-organized structures referred to as *systems*. The world, both around and inside us, consists of many such systems. Most systems of interest are not fully isolated, but interact with each other. Their interactions may be due to fundamental physical forces, such as gravity or electromagnetism, collisions, or exchange of energy or matter.

In classical mechanics, the study of motion of bodies induced by external or internal forces is called *dynamics*, a word which originates from the Greek word δύναμις, meaning "power." In a more general context, we understand the term *dynamical* to be equivalent with *time-dependent*. Therefore, a *dynamical system* is a system that evolves over time. Hereby, we tacitly adopt the Newtonian concept of a globally defined time, i.e., we assume that the system variables are functions of time, treated as a universal parameter. In the context of mechanics, the system variables are typically positions and velocities of idealized mass points and, in the simplest case, the evolution of the system variables is determined by Newton's laws. However, in the following we will not restrict ourselves to mechanical systems, but will allow for more general system variables: for instance, concentrations of chemicals, intensity of a light beam, or temperature at a given point. Note that the system variables can be functions of both space and time.

Unless stated otherwise, we will focus on the case of *deterministic* systems, i.e., dynamical systems that are not influenced by noise. Mathematically, a dynamical system can be described by either differential or difference equations. In the former case, time flows continuously and the system is called a *continuous system* or *flow*. In the latter case, time changes discretely, and the system is known as a *discrete system*. In the following, we will describe the main features of dynamical systems by referring to popular examples.

1.1.1 Linear Dynamical Systems

Dynamical systems whose variables are linked by linear functions are called *linear systems*. The temporal evolution of a continuous linear system characterized by n system variables is generated by a set of n ordinary differential equations

$$
\begin{aligned}
\dot{x}_1 &= a_{11}x_1 + a_{12}x_2 + \cdots + a_{1n}x_n, \\
\dot{x}_2 &= a_{21}x_1 + a_{22}x_2 + \cdots + a_{2n}x_n, \\
&\;\;\vdots \\
\dot{x}_n &= a_{n1}x_1 + a_{n2}x_2 + \cdots + a_{nn}x_n,
\end{aligned}
\tag{1.1}
$$

where $x_i = x_i(t)$ are the time-dependent system variables, $\dot{x}_i \equiv dx_i/dt$ are their time derivatives, and a_{ij} are constant coefficients.

In vector form, Equation 1.1 can be written as

$$
\dot{\mathbf{x}}(t) = \mathbf{A}\mathbf{x}(t),
\tag{1.2}
$$

where $\mathbf{x} = (x_1, x_2, \ldots, x_n)$ is an n-dimensional vector, and $\mathbf{A}$ is a constant matrix.

When the system dynamics are defined in terms of discrete times, i.e., when the current state of the system is iteratively determined by the previous one, the dynamics are instead described by difference equations, or *iterative maps*. The variables exhibit a mapping form, as $\mathbf{x}$ varies in discrete steps:

$$
\mathbf{x}_{i+1} = \mathbf{A}\mathbf{x}_i.
\tag{1.3}
$$

Linear systems can be solved exactly. The solution of Equation 1.2 has an exponential form and can be found using the set of eigenvalues λ, given by the determinant equation

$$\det(A - \lambda \mathbf{I}) = 0, \tag{1.4}$$

where $\mathbf{I}$ is the identity matrix, and the eigenvectors $\mathbf{v_i}$ satisfy the equation

$$\mathbf{Av_i} = \lambda_i \mathbf{v_i}. \tag{1.5}$$

The eigenvalues λ represent powers of the exponential components of the solution, and the eigenvectors are their coefficients.

Pure linear systems, however, do not exist in nature. Like a point mass, they are just mathematical approximations. The dynamics of linear systems, indeed, are not rich enough to describe the most commonly observed behaviors, such as periodic oscillations, bifurcations, synchronization, and chaos. The asymptotic state of a bounded linear system, reached for $t \to \infty$, is only a steady state, i.e., a fixed equilibrium point that can be either stable or unstable. The stability properties of dynamical systems will be described in Section 1.4.

1.1.2 Nonlinear Dynamical Systems

If a system is characterized by variables that depend nonlinearly on each other, the motion can become very complex. Such systems are called *nonlinear dynamical systems*. Mathematically, a nonlinear continuous dynamical system is described by

$$\begin{aligned} \dot{x}_1 &= F_1(x_1, x_2, \ldots, x_n), \\ \dot{x}_2 &= F_2(x_1, x_2, \ldots, x_n), \\ &\vdots \\ \dot{x}_n &= F_n(x_1, x_2, \ldots, x_n), \end{aligned} \tag{1.6}$$

where F_i are functions that couple the variables among them. If at least one of these functions is nonlinear, the system in Equation 1.6 is said to be nonlinear.

In vector form, a nonlinear dynamical system can be described as

$$\dot{\mathbf{x}}(t) = \mathbf{F}(\mathbf{x}(t)), \tag{1.7}$$

where $\mathbf{F} = (F_1, F_2, \ldots, F_n)$ is a vector function: $\mathbb{R}^n \to \mathbb{R}^n$.

The time evolution of the system describes a *trajectory*, or *orbit*, in the Euclidean space of the n variables $\mathbf{x} \in \mathbb{R}^n$, or *phase space*. Each point in the phase space represents a unique state of the system. In the case of three-dimensional systems, one can visualize directly the trajectory in three coordinates (x_1, x_2, x_3), while for systems with $n > 3$, visualization of the orbit is only possible by means of projections of the phase space on planes (or hyperplanes) of two or three of the system's variables.

Since at any given time the system state is described by a vector defined by the functions and parameters in Equation 1.7, the system evolution is *deterministic*.

In other words, for any fixed initial condition $\mathbf{x}_0 \in \mathbb{R}^n$, the system always follows a unique path, which therefore can never intersect the paths originating from different initial conditions. However, it has to be remarked that a strictly deterministic system is only a theoretical idealization, because random components or fluctuations are always present in nature, and are sometimes accounted for via perturbation theory. Furthermore, in practice, exact knowledge about the future state of a system is restricted by the precision with which the initial state can be measured, especially for chaotic systems characterized by a strong dependence on initial conditions.

1.1.3 Autonomous and Nonautonomous Systems

A dynamical system that contains a time-dependent function is called *nonautonomous*; otherwise, the system is said to be *autonomous*. Every nonautonomous system can be transformed into an autonomous system by adding an additional degree of freedom proportional to the time. As an example, let us consider such a transformation when applied to a CO_2 laser model. Under loss modulation, this laser represents a nonautonomous (or driven) system described as (Chizhevsky et al. 1997; Pisarchik and Corbalán 1999)

$$\dot{x} = \tau^{-1}x\,(y - k_0 - k_m \sin(2\pi f_m t))\,,$$
$$\dot{y} = (y_0 - y)\gamma - yx\,. \tag{1.8}$$

In these equations, x is proportional to the radiation density, y and y_0 stand for the gain and the unsaturated gain in the active medium, τ is the light half round-trip time in the laser cavity, γ is the gain decay rate, k_0 is the constant portion of the losses, and k_m and f_m are the modulation amplitude and frequency. The system is nonlinear because of the yx coupling term.

Introducing the additional variable $z = 2\pi f_m t$, one can convert the two-dimensional nonautonomous system in Equation 1.8 into the three-dimensional autonomous system:

$$\dot{x} = \tau^{-1}x(y - k_0 - k_m \sin z),$$
$$\dot{y} = (y_0 - y)\gamma - yx, \tag{1.9}$$
$$\dot{z} = 2\pi f_m.$$

It is clear that, by generalization of this procedure, any nth-order nonautonomous system can be transformed into an $(n + 1)$-dimensional autonomous system.

1.1.4 Conservative and Dissipative Systems

A dynamical system is said to be conservative (dissipative) if a unitary volume of initial conditions produces orbits whose images in time are contained within a constant (contracting) volume of the phase space. Although all real dynamical systems are dissipative, quantum mechanics mostly deals with conservative, or Hamiltonian, systems. The most notable examples of conservative systems are undamped pendula, sets of point masses interacting under Newton's gravitational force, and nonrelativistic charged particles in an electromagnetic field.

Dissipation arises from any kind of loss, quite often due to internal friction. In dissipative dynamical systems, the potential, or energy, goes from an initial form to a final asymptotic one. After a sufficiently long period – known as *transient* time – has elapsed, the trajectory of the dissipative system is found in a subset of the phase space that is said to be the system's *attractor*. As the energy in conservative systems is preserved, they do not have attractors.

A typical example of a dissipative linear system is a damped harmonic oscillator, described by

$$m\ddot{x} + c\dot{x} + kx = 0, \tag{1.10}$$

where m is the mass, c is the viscous damping coefficient, and k is the elastic constant. These coefficients define the undamped angular frequency

$$\omega_0 = \sqrt{k/m} \tag{1.11}$$

and the damping ratio

$$\zeta = \frac{c}{2\sqrt{mk}}. \tag{1.12}$$

When the new variables $x_1 = x$ and $x_2 = \dot{x}$ are introduced, Equation 1.10 generates a system of two first-order differential equations:

$$\begin{aligned}\dot{x}_1 &= x_2, \\ \dot{x}_2 &= -(c/m)x_2 - \omega_0^2 x_1.\end{aligned} \tag{1.13}$$

The damping ratio ζ determines the transient behavior of the oscillator. If $\zeta = 0$, the oscillator is undamped, the system is conservative, and the solutions are oscillations that continue indefinitely with frequency ω_0. If $0 < \zeta < 1$, the oscillator is *underdamped*, and its oscillations have a frequency $\omega = \omega_0\sqrt{1 - \zeta^2}$. Otherwise, the system returns to its equilibrium without oscillating, and the oscillator is known as either *overdamped* (if $\zeta > 1$) or *critically damped* (if $\zeta = 1$).

All systems considered in this book behave like undamped or underdamped oscillators, since speaking of synchronization for fixed points is meaningless.

1.2 Chaotic Systems

Nonlinear differential equations are very difficult (or even impossible) to solve analytically, and until computer simulations became possible, chaotic solutions could not be calculated (Lorenz 1963).

By chaotic solutions, we generally denote those trajectories that have a critical dependence on the initial conditions. This means that if one considers any two trajectories originating from two nearby initial conditions (whose Euclidean distance in phase space is arbitrarily small), these trajectories *exponentially* separate in time, i.e., the distance between the two actual states of the system grows exponentially over time.

To discuss the main concepts of chaotic dynamics, let us consider a three-dimensional ($n = 3$) nonlinear Rössler oscillator, described by the system (Rössler 1977)

$$\dot{x} = -\omega y - z,$$
$$\dot{y} = \omega x + ay, \qquad\qquad (1.14)$$
$$\dot{z} = b + z(x - c).$$

This oscillator is often used for studying synchronization because its natural frequency ω is directly included in the equations as a parameter.

In spite of its apparent simplicity, the system in Equation 1.14 exhibits very rich dynamics. For a large set of parameters, such as, for example, $a = 0.16$, $b = 0.1$, $c = 8.5$, and $\omega = 1$, the motion of the system Equation 1.14 is chaotic.

The dynamics of a nonlinear system can be visualized and characterized by the following tools: (i) time series, (ii) phase-space portrait, and (iii) power spectrum. Let us consider each of these tools separately.

1.2.1 Time Series

Time series describe the temporal evolution of a system variable. Figure 1.1 shows the time series of all three variables of Equation 1.14 in the chaotic state given by the parameter values defined above. Although the oscillations of each variable are irregular, they are correlated due to their functional dependence in Equation 1.14, even if we cannot see it at a first glance.

Time series have proven to be a good tool for synchronization analysis, to extract meaningful statistics and other data characteristics. Time series analysis can be carried out either in the time domain or in the frequency domain. The former includes cross-correlation analyses that are frequently used to quantify synchronization of coupled systems (see Section 2.3). The latter includes spectral analysis, which is described below. Time series analysis can also be used to reconstruct attractors from experimental data.

1.2.2 Phase Space

The concept of phase space was developed in the nineteenth century thanks to the contributions of Ludwig Boltzmann, Henri Poincaré, James Maxwell, and Josiah Gibbs to statistical mechanics and Hamiltonian mechanics. In nonlinear dynamics,

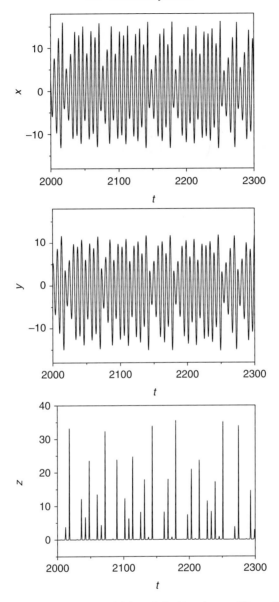

Figure 1.1 Time series of three variables of the Rössler oscillator, Equation 1.14, in the chaotic regime for $a = 0.16$, $b = 0.1$, $c = 8.5$, and $\omega = 1$.

the phase space is a space whose coordinates correspond to the system variables. The system trajectory in the phase space represents all possible states during an infinite time evolution. The phase space dimension is equal to the number of system variables. The phase space trajectory of the chaotic Rössler oscillator is shown in Figure 1.2.

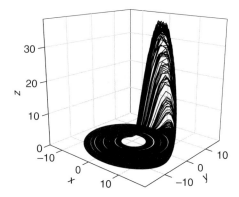

Figure 1.2 The chaotic trajectory of the Rössler oscillator for $a = 0.16$, $b = 0.1$, $c = 8.5$, and $\omega = 1$, contained within a certain region of the phase space (x, y, z).

When the phase space has a very large dimension, the trajectory of a system is often visualized by projections onto the subspace corresponding to two or three variables. These projections are called *phase portraits*.

1.2.3 Power Spectrum

Another way to visualize the dynamics of a system is via a *power* or *frequency spectrum* of one of the system variables. The power spectrum can be obtained by means of the *Fourier transform* of the time series. The Fourier transform $\mathcal{F}$, named after Joseph Fourier (1768–1830), is a mathematical transformation employed to transform a signal from a time domain to a frequency domain. A reverse operation $\mathcal{F}^{-1}$ is also possible. Mathematically, the direct and inverse Fourier transforms are defined as

$$\mathcal{F}(x(t)) \equiv X(\omega) = \frac{1}{\sqrt{2\pi}} \int_{-\infty}^{\infty} x(t) e^{-i\omega t} dt , \tag{1.15}$$

$$\mathcal{F}^{-1}(X(\omega)) \equiv x(t) = \frac{1}{\sqrt{2\pi}} \int_{-\infty}^{\infty} X(\omega) e^{i\omega t} d\omega . \tag{1.16}$$

These integrals exist if three conditions are met, namely:

(i) $x(t)$ is piecewise continuous;
(ii) $x(t)$ is piecewise differentiable;
(iii) $x(t)$ is absolutely integrable, i.e., $\int_{-\infty}^{\infty} |x(t)| dt$ is finite.

Then, the power spectrum is defined as

$$S(\omega) = X^*(\omega) X(\omega) = |X(\omega)|^2 , \tag{1.17}$$

where X^* is the complex conjugate of X.

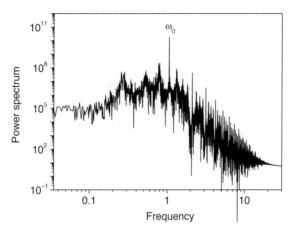

Figure 1.3 Power spectrum of the *y* variable of a Rössler oscillator with the parameter choice $a = 0.16, b = 0.1, c = 8.5$, and $\omega = 1$. Notice that the spectrum of the chaotic signal is continuous, and has a peak at the dominant frequency ω_0.

Figure 1.3 shows the power spectrum of the *y* variable of a Rössler oscillator with the usual parameters, obtained from the time series of Figure 1.1. The power spectrum of the chaotic regime has a maximum at a dominant frequency ω_0. Due to nonlinearity, ω_0 differs from ω by a small amount.

1.3 Attractors

Differential and difference nonlinear equations that describe dynamical systems give rise to many types of solutions, both stable and unstable. Stable solutions encountered in nonlinear dynamics are *attractors*: asymptotically stable volumes of the phase space toward which a system evolves, when starting from a set of initial conditions known as the attractor's *basin of attraction*.

1.3.1 Types of Attractors

Attractors are therefore portions (or subsets) of the phase space. The simplest possible attractor is a stable fixed point, which can be found even in linear systems (Section 1.1). Nonlinearity, however, allows for more complex and interesting attractors. They come in different geometric shapes in phase space, such as limit cycles (periodic orbits), toroids, and miscellaneous manifolds, and may even have a fractal structure (strange attractors).

A single nonlinear dynamical system can exhibit different attractors, depending on the values chosen for its parameters. Let us illustrate the case with the Rössler oscillator of Equation 1.14. In Figure 1.4 we show three different attractors in the phase space. Each of these attractors is determined by the parameter *c*, while keeping the other parameters unchanged.

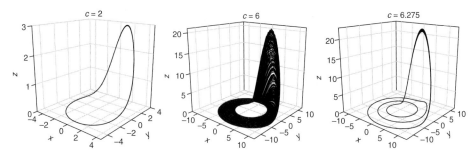

Figure 1.4 Types of attractors of a Rössler oscillator for different values of the parameter c. (a) A period-one orbit is generated for $c = 2$; (b) a chaotic attractor is generated for $c = 6$; and (c) a period-three orbit is generated for $c = 6.275$. The other parameters are $a = 0.16$, $b = 0.1$, and $\omega = 1$.

Some nonlinear systems even allow coexistence of attractors. While all parameters remain constant, the attractor changes according to the initial conditions. Such systems are called *multistable*. Multistability can be revealed not only by changing initial conditions, but also by varying a system parameter back and forth (continuation method; see Seydel 1988), or by adding noise that converts the multistable system to a metastable one (noise-induced multistate intermittency; see Pisarchik et al. 2012a).

1.3.2 Basins of Attraction and Poincaré Maps

The basin of attraction of an attractor is the set of initial conditions that lead the asymptotic trajectory of the system to the attracted state. For our example of Equation 1.14, the basin of attraction of the chaotic attractor has a very sophisticated structure in the three-dimensional phase space.

A visualization of the attractor can be given in a two-dimensional space by plotting the intersection points of trajectories with a plane corresponding to a certain (fixed) value of one of the system variables (say, z). Such a plane is called the *Poincaré section*, named after French mathematician Henri Poincaré (1854–1912). The attractor can be visualized on the Poincaré section by plotting the value of the function each time it passes through it in a specific direction. This map is called the *Poincaré map*, and it is a lower-dimensional subspace transversal to the system flow in phase space.

The Poincaré section is a very useful tool for the analysis of nonlinear dynamics, as well as for revealing the attractor's structure in spaces where dimensions are reduced by one. Indeed, given a flow, a Poincaré map can always be constructed, composed by intersection points $\mathbf{x_j}(t_j)$ on the Poincaré section in one direction at discrete times t_j. This procedure converts the flow Equation 1.7 to the map $\mathbf{x_{j+1}}(t_{j+1}) = \mathbf{P}(\mathbf{x}_j(t_j))$.

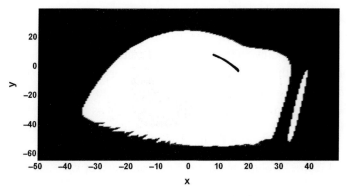

Figure 1.5 Poincaré map of the chaotic attractor (black dots in the central part) with its basin of attraction (white region) for the Rössler oscillator of Equation 1.14 and parameters $a = 0.16$, $b = 0.1$, $c = 8.5$, and $\omega = 1$, on the Poincaré section obtained for $z = 5$.

Another way of visualizing the attractors is by means of the stroboscopic map, which is made by plotting a single scalar variable $x_n(t_j)$ at local maxima (or minima) of the time series versus a control parameter.

Figure 1.5 shows the Poincaré map, and the basin of attraction on the Poincaré section at $z = 5$, for the Rössler oscillator of Equation 1.14. It seems reasonable that if an initial condition asymptotes to a particular attractor, then nearby trajectories would also asymptote to the same state. However, this does not always happen. Instead, in coupled chaotic systems, one may have that arbitrarily close points of the phase space belong to basins of attraction of different states. Such basins were discovered by James Alexander and coworkers (1992), and are referred to as *riddled basins* of attraction.

1.4 Stability of Dynamical Systems

The information about the stability of the solutions of a dynamical system is vital for understanding the mechanisms responsible for the system behavior. In the following, we briefly concentrate on linear stability analysis, Lyapunov exponents, and bifurcations.

1.4.1 Linear Stability Analysis

Among the various stability properties, *linear* or *exponential* stability occupies a particular place, because it indicates *local* stability of an equilibrium point of a dynamical system.[1]

---

[1] It may be worth noting that linear and exponential stability are one and the same, as one can go back and forth using a logarithmic scale.

Linear stability analysis is performed by *linearizing* differential or difference equations, in order to find the linear approximation of a nonlinear function in the proximity of a given point. In the case of differential equations, the linearization of the function $\mathbf{F}(\mathbf{x})$ is the first-order term of its Taylor expansion around the specific equilibrium point $\mathbf{x_0}$ whose stability properties have to be assessed.

In its linearized form, Equation 1.7 can be written as

$$\dot{\mathbf{x}} = \mathbf{F}(\mathbf{x_0}) + J_{\mathbf{F}}(\mathbf{x_0})(\mathbf{x} - \mathbf{x_0}), \qquad (1.18)$$

where $J_{\mathbf{F}}(\mathbf{x_0})$ is the Jacobian matrix of $\mathbf{F}(\mathbf{x})$ evaluated at $\mathbf{x_0}$:

$$J_{\mathbf{F}}(\mathbf{x_0}) = \begin{pmatrix} \frac{\partial F_1}{\partial x_1} & \cdots & \frac{\partial F_1}{\partial x_n} \\ \vdots & \ddots & \vdots \\ \frac{\partial F_n}{\partial x_1} & \cdots & \frac{\partial F_n}{\partial x_n} \end{pmatrix}_{\mathbf{x} = \mathbf{x_0}}. \qquad (1.19)$$

The eigenvalues of the Jacobian matrix allow one to determine the nature of the equilibrium point $\mathbf{x_0}$. Specifically, near the equilibrium point, the system is *stable* if the real parts of all the eigenvalues are negative, and *unstable* otherwise. In other words, a single eigenvalue real part greater than 0 is enough to break stability. If the largest real part of the eigenvalues is equal to 0, the point is said to be *marginally stable*.

An equilibrium point is called *hyperbolic* if all eigenvalues have a nonzero real part. In three-dimensional continuous nonlinear systems, there are eight types of hyperbolic equilibrium points, corresponding to the eight possible sign combinations of the real parts of the eigenvalues.

A nonhyperbolic equilibrium point has one or more eigenvalues with a zero real part. There are nineteen such types of equilibrium points in three-dimensional flows. The stability of those systems that do not have an eigenvalue with a positive real part cannot be determined from the eigenvalues and requires more sophisticated nonlinear analysis.

Let us again consider the example of the Rössler oscillator in Equation 1.14. Setting $\dot{x} = \dot{y} = \dot{z} = 0$, we obtain two equilibrium points:

$$\begin{aligned} x_{1,2}^0 &= -\frac{-c \pm \sqrt{c^2 - 4ab/\omega^2}}{2}, \\ y_{1,2}^0 &= \omega \frac{-c \pm \sqrt{c^2 - 4ab/\omega^2}}{2a}, \\ z_{1,2}^0 &= -\omega^2 \frac{-c \pm \sqrt{c^2 - 4ab/\omega^2}}{2a}. \end{aligned} \qquad (1.20)$$

For the chaotic attractor shown in Figure 1.2 the fixed points are $\mathbf{x_1^0} = (x_1^0, y_1^0, z_1^0) = (1.88 \times 10^{-3}, -1.18 \times 10^{-2}, 1.18 \times 10^{-2})$ and $\mathbf{x_2^0} = (x_2^0, y_2^0, z_2^0)$

$= (8.50, -53.11, 53.11)$. One of these fixed points resides at the center of the attractor loop and the other lies outside the attractor.

The linear approximation of Equation 1.14 near these points yields the following Jacobian matrix:

$$J\left(\mathbf{x}_{1,2}^{0}\right) = \begin{pmatrix} 0 & -\omega & -1 \\ \omega & a & 0 \\ z_{1,2}^{0} & 0 & x_{1,2}^{0} - c \end{pmatrix}. \tag{1.21}$$

The stability of each of these fixed points can be assessed by determining the respective eigenvalues of the Jacobian matrix, by solving the characteristic equation

$$\det\left(J\left(\mathbf{x}_{1,2}^{0}\right) - \lambda \mathbf{I}\right) = \begin{vmatrix} -\lambda & -\omega & -1 \\ \omega & a - \lambda & 0 \\ z_{1,2}^{0} & 0 & x_{1,2}^{0} - c - \lambda \end{vmatrix} = 0. \tag{1.22}$$

This yields

$$-\lambda^{3} + \lambda^{2}(a + x_{1,2}^{0} - c) + \lambda(ac - ax_{1,2}^{0} - \omega^{2} - z_{1,2}^{0}) + \omega^{2}(x_{1,2}^{0} - c) + az_{1,2}^{0} = 0, \tag{1.23}$$

from which it follows that $\mathbf{x}_{1}^{0}$ has eigenvalues $\lambda_{1,2} = 7.93 \times 10^{-2} \pm 0.997i$ and $\lambda_{3} = -8.50$, and $\mathbf{x}_{2}^{0}$ has eigenvalues $\lambda_{1,2} = 0.16 \pm 7.36i$ and $\lambda_{3} = -0.16$. Each fixed point has two complex conjugated eigenvalues $\lambda_{1,2} = \Re(\lambda) \pm i\nu$ and one real eigenvalue λ_{3}. The ν in the complex conjugate eigenvalues is the eigenfrequency of the corresponding oscillation mode.

The magnitude of the real part characterizes the speed of attraction or repulsion along the corresponding eigenvector, with a negative real part indicating attraction and a positive real part repulsion. Since both fixed points have negative real and two complex conjugated eigenvalues with positive real parts, they are *unstable foci*. The trajectory approaches these points without oscillations along a stable manifold and leaves it along an unstable manifold by making an untwist spiral with eigenfrequency ν.

An equilibrium point is known as a *saddle* if it is a hyperbolic equilibrium point with at least one eigenvalue with positive real part and at least one with negative real part. The saddle point connects stable and unstable manifolds. The occurrence of all these conditions implies that both fixed points in our example are saddles.

1.4.2 Lyapunov Exponents

Named after the Russian mathematician and physicist Aleksandr Lyapunov (1857–1918), the Lyapunov exponents describe the rate at which nearby trajectories diverge from (or converge to) each other. In other words, they are a measure of the system sensitivity to a small change in initial conditions, and fully characterize a chaotic system.

Consider a system evolving from two slightly different initial conditions $\mathbf{x_0}$ and $\mathbf{x_0} + \varepsilon(0)$. In order to find the time evolution of $\varepsilon(t)$, we linearize Equation 1.7 so that

$$\dot{\varepsilon}(t) = J_F(\mathbf{x}(t))\varepsilon(t) . \tag{1.24}$$

The Lyapunov exponent is defined as

$$\lambda(\mathbf{x_0}, \varepsilon) = \lim_{t\to\infty} \left(\frac{1}{t} \log \frac{|\varepsilon(t)|}{|\varepsilon(0)|} \right) . \tag{1.25}$$

The number of Lyapunov exponents of a dynamical system is equal to the number of its variables, and the complete spectrum of the Lyapunov exponents fully characterizes the system stability.

In some cases, the knowledge of the largest Lyapunov exponent (the first, if they are sorted in decreasing order) is enough to identify the system behavior. A nonpositive largest Lyapunov exponent means that two nearby trajectories will always converge and the system is not chaotic. A positive largest exponent indicates that nearby trajectories diverge, resulting in a chaotic motion.

If a system has more than one positive Lyapunov exponents, the system is said to be *hyperchaotic*.

1.4.3 Bifurcations

In general, a small variation in some parameter produces small, continuous changes in the system output, so that the system is said to be *structurally stable*. However, for some specific parameter values, a small variation can induce a strong qualitative change in the solution. Such a behavior is called a *bifurcation*, and the system is said to be *structurally unstable* for these parameter values.

The parameter value at which a bifurcation occurs is called *critical*, or a *bifurcation point*.[2] At this point, the attractor changes its stability, and this may result in a change of the attractor type: for instance, a fixed point may be converted into a limit cycle or chaos.

Depending on the changes that the system undergoes, different bifurcation types can be observed. In one-dimensional dynamical systems there are four basic types of bifurcation: saddle-node, transcritical, and pitchfork bifurcations (super- and subcritical). In a *saddle-node bifurcation*, stable and unstable fixed points collide

---

[2] In sociology and climatology a sudden change of the system dynamics is often called a *tipping point* after Gladwell (2000), who introduced this notion to define a bifurcation in time when a group of people dramatically changes its behavior. A climate tipping point means a point when global climate changes from one stable state to another (Kopp et al. 2016).

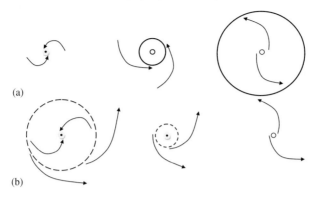

Figure 1.6 Phase space trajectories before, at and after (a) supercritical and (b) subcritical Andronov–Hopf bifurcation. Increasing control parameter from left to right, the phase space topography changes (a) from a globally stable fixed point through a birth of an unstable fixed point and a stable limit cycle with increasing amplitude or (b) from a stable fixed point through a birth of an unstable limit cycle with decreasing amplitude to an unstable fixed point.

and annihilate; in a *transcritical bifurcation*, stable and unstable fixed points exchange their stability; in a *supercritical pitchfork bifurcation*, a stable fixed point becomes unstable and two new stable fixed points arise; and in a *subcritical pitch-fork bifurcation*, one stable and two unstable fixed points collide and the original attractor become a repeller.

The most important bifurcation in a two-dimensional dynamical system is the *Andronov–Hopf bifurcation*, which can be either subcritical or supercritical. This bifurcation can be thought of as a critical point where a stable steady state trans-forms into a periodic orbit (supercritical), or a stable fixed point transforms into an unstable limit cycle (subcritical). The schematic phase space trajectory in the vicinity of this type of bifurcation is illustrated in Figure 1.6.

Very often, when a control parameter is varied and a bifurcation appears at some critical value, this is a signature of the existence of a sequence of new bifurcations at higher values of the control parameter, with each new attractor appearing in the bifurcation chain usually being more complex than the previous one. When starting from a fixed point and ending with a chaotic attractor, the sequence of bifurcations is called a *road* (or *route*) *to chaos*.

The number of possible routes to chaos is yet unknown, but it has been observed that some of them appear very often, and for this reason they are called *scenarios*. A good tool for visualization of bifurcation scenarios is a *bifurcation diagram*. The bifurcation diagram is a graphical representation of a system's behavior with respect to a control parameter. The bifurcation diagrams can be created by using either a Poincaré map or a stroboscopic map. The most common scenarios to chaos are the Feigenbaum scenario (period-doubling route), the Ruelle–Takens–Newhouse scenario (quasiperiodicity route), and the Pomeau–Manneville scenario (intermittency route) (Eckmann 1981). We will give a brief description of each in the following sections.

Feigenbaum Scenario (Period-Doubling Route to Chaos)

One of the most common routes to chaos in dynamical systems is the one occurring through a sequence of subharmonic bifurcations which appear in the following way

$$T \rightarrow \mu_1 \rightarrow 2T \rightarrow \mu_2 \rightarrow 4T \rightarrow \mu_3 \rightarrow 8T \rightarrow \mu_4 \rightarrow \ldots \mu_\infty \rightarrow \text{chaos},$$

where $2mT$ ($m = 0, 1, 2, \ldots$) represents the period of a limit cycle and $\mu_n(n = 1, 2, 3, \ldots)$ denote the critical values for the control parameter μ. This sequence is known as the Feigenbaum scenario, after American physicist Mitchell Feigenbaum, who was among the first to describe its universal properties. The Feigenbaum route to chaos has been observed in a variety of systems, including loss-modulated CO_2 lasers (Arecchi et al. 1982).

A typical bifurcation diagram demonstrating a period-doubling route to chaos is shown in Figure 1.7(a). The diagram displays the dependence of the peak values (local minima) of the variable y of the Rössler oscillator as a function of the parameter c.

The corresponding Lyapunov exponents are illustrated in Figure 1.7(b). The bifurcation diagram represents a cascade of period-doubling bifurcations: period 1, period 2, period 4, period 8, etc., terminating with chaos at $c \approx 4.8$, where the largest Lyapunov exponent (λ_x) becomes positive. For higher values of c, windows of a periodic motion of periods 3, 4, and 5 appear in the bifurcation diagram. Within these windows, all Lyapunov exponents return to be nonpositive. At the bifurcation points, the two largest Lyapunov exponents are zero, meaning that the system is periodic.

Ruelle–Takens–Newhouse Scenario (Quasiperiodicity Route to Chaos)

This second scenario consists of a sequence of three bifurcations: Andronov–Hopf bifurcation (μ_1) and two torus bifurcations (μ_2 and μ_3)

$$FP \rightarrow \mu_1 \rightarrow T \rightarrow \mu_2 \rightarrow T^2 \rightarrow \mu_3 \rightarrow T^3(\text{chaos}).$$

The typical bifurcation diagram for this scenario is shown in Figure 1.8, which shows the peak intensity of a semiconductor laser with feedback using the feedback strength as a control parameter. As seen from the diagram, the fixed point attractor (FP) (or stable steady state) is converted into a periodic orbit T in the Andronov–Hopf bifurcation at $\mu = \mu_1$.

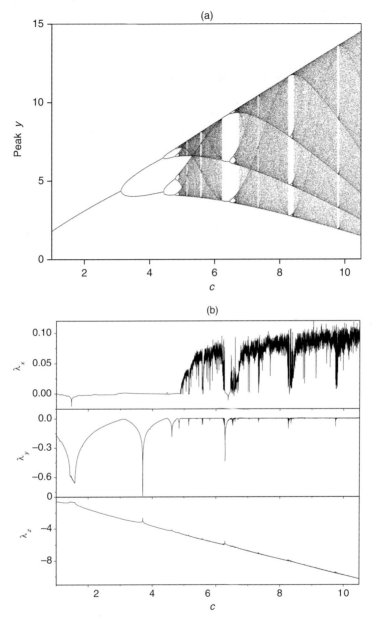

Figure 1.7 (a) A typical bifurcation diagram corresponding to a period-doubling route to chaos, and (b) the three Lyapunov exponents of the Rössler oscillator of Equation 1.14 as a function of the parameter c, which is taken here as the control parameter. The other parameters are $a = 0.16$, $b = 0.1$, and $\omega = 1$.

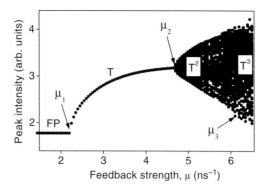

Figure 1.8 Typical bifurcation diagram corresponding to a quasiperiodic route to chaos. The diagram has been obtained for a semiconductor laser with feedback, using the feedback strength μ as the control parameter. The fixed point attractor (FT) is converted into a periodic orbit T, which transforms into a torus T^2 and a chaotic attractor T^3.

Then, at the torus bifurcation (sometimes called the secondary Hopf bifurcation, $\mu = \mu_2$) the periodic orbit T transforms into a torus T^2, which entails a quasi-periodic behavior with two incommensurate frequencies, and finally at the second torus bifurcation ($\mu = \mu_3$) a new independent frequency appears, so that in principle a T^3 attractor (three-dimensional torus or hypertorus) is expected, even though in many cases it is unstable towards some kinds of fluctuations and becomes chaotic.

The Ruelle–Takens–Newhouse scenario can be identified by observing the power spectrum of a dynamical variable, such as those shown in Figure 1.9. In the hyper-torus bifurcation μ_3, the quasiperiodic spectrum of panel (a) is converted into the broadband chaotic spectrum of panel (b), where some peaks are usually still apparent, indicating that the previous periodic evolution has not completely disappeared.

Pomeau–Manneville Scenario (Intermittency Route to Chaos)

The intermittency route to chaos is characterized by short, irregular, turbulent bursts, interrupting a nearly regular, laminar motion (Figure 1.10). The duration of the turbulent phases is fairly regular and weakly dependent on the control parameter μ, but the mean duration of the laminar phase decreases as μ increases beyond its critical value, and eventually the laminar phase disappears altogether. Hence, only one bifurcation point is associated with the intermittency route to chaos.

Among many types of intermittency the most notable are type-I, type-II, and type-III of Pomeau–Manneville intermittency (Pomeau and Manneville 1980), on-off (Platt et al. 1993), and crisis-induced intermittency (Fujisaka et al. 1983).

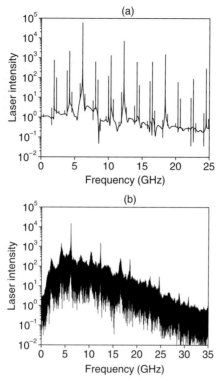

Figure 1.9 Power spectra of laser intensity corresponding to (a) a quasiperiodic and (b) a chaotic behavior. Chaos is characterized by a continuous, broadband spectrum.

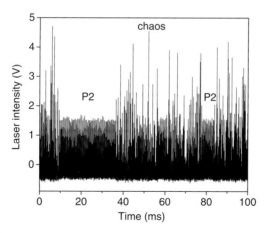

Figure 1.10 Time evolution of the laser intensity, during an on-off intermittency phenomenon in a fiber laser with pump modulation. The period-two (P2) regime alternates (at random times) with the coexisting chaotic regime.

A particular type of intermittency depends on the type of bifurcation at the critical point. The type-I and on-off intermittency are associated with saddle-node bifurcations, the type-II and type-III intermittency with Andronov–Hopf bifurcation and inverse period-doubling bifurcation, respectively, and crisis-induced intermittency with crisis of chaotic attractors when two (or more) chaotic attractors simultaneously collide with a periodic orbit (or orbits).

Quantitatively, intermittency exhibits characteristic interburst interval (laminar phase) statistics. The mechanism for on-off intermittency relies on the time-dependent forcing (stochastic or periodic) of a bifurcation parameter through a bifurcation point, while the system switches between two or more unstable states, which are stable without external forcing. Conversely, in Pomeau–Manneville intermittency and crisis-induced intermittency the parameters are static.

To conclude, we note that this list of possible routes to chaos is not exhaustive. A number of other scenarios have been identified, such as, for example, via a cascade of homoclinic bifurcations (Arnéodo et al. 1981), in crisis (Ott 2002) as well as other scenarios (Newhouse et al. 1983; MacKay and Tresser 1986). For multistable systems, the structure of bifurcation diagrams can be even more complex, exhibiting branches of different coexisting attractors.

2

Low-Dimensional Systems

In this chapter we briefly describe the most important synchronization states that emerge when low-dimensional systems are coupled (or subjected to an external common noise or force), such as complete synchronization, phase synchronization, lag and anticipated synchronization, and generalized synchronization.

Various routes from asynchronous state to synchrony are explored, together with their dependence on coupling strength and configuration. These pave the way for the discussion, in the next chapter, of multistable system, effects of noise, and important applications to neural systems and secure communications.

Once again, for completeness, we refer interested readers to the following bibliography (Pikovsky et al. 2001; Boccaletti et al. 2002; González-Miranda 2004; Boccaletti 2008; Luo 2013), where they will find complementary (and more exhaustive) descriptions of the same phenomena, as well as in-depth details on each of the synchronization states mentioned.

2.1 A Brief History of Synchronization

The first recorded use of the verb "synchronize" is found in the annals of 1624, with the meaning of "to occur at the same time." This word originated from Greek and it's composed by the prefix $\sigma \acute{\upsilon}\nu$-, meaning "together," and the word $\chi\rho\acute{o}\nu o\varsigma$, "time" (Pickett et al. 2011).

The transitive sense of "make something synchronous" was first recorded in 1806. Later, in 1879, this verb was used in relation to timepieces and, in 1950, used for the first time to indicate the occurrence of "synchronized swimming."

Synchronization is commonly understood as a collective state of coupled systems. In general, it indicates the existence of some relation between functionals of different processes due to their interactions (Boccaletti et al. 2002). Synchronization is also a process during which coupled oscillatory systems adjust their individual frequencies in an organized fashion, and it is related to self-control or self-organization of matter.

Figure 2.1 Examples of systems featuring collective synchronous motion: flocks of birds, coordinating aircrafts, orchestras, schools of fishes.

Synchronized behavior is extremely important in science and engineering, and has numerous applications in many fields, from mechanics and electronics to physics, chemistry, biology and even economics. Synchronization is furthermore ubiquitous in nature and human life. Some examples of synchronous motion that are observed in real-world systems are illustrated in Figure 2.1. In all of them, each element of the system coordinates its behavior according to the motion of its neighbors and/or to an external force. For example, a symphony orchestra is synchronized by the conductor, and a school of fish changes its shape when attacked by sharks.

Experimental studies on synchronization of periodic oscillators were started by the Dutch mathematician and physicist Christiaan Huygens (1629–95), who worked with pendulum clocks as far back as the seventeenth century (1913).

The first nonlinear theory was created only in the beginning of the twentieth century by another Dutch physicist, Balthasar van der Pol (1889–1959), while he was working at Philips. Van der Pol explored electrical circuits employing vacuum tubes and found that they exhibited stable oscillations, now known as limit cycles. When these circuits were driven by a periodic signal whose frequency was close to that of the limit cycle, their oscillation frequencies became entrained by the external drive, i.e., they shifted to the frequency of the driving signal (Van der Pol and Van der Mark 1927). This discovery had a great deal of practical importance because vacuum tubes were, at that time, the basic elements of radio communication systems.

Later, the synchronization theory of periodic oscillations was developed by Russian mathematicians (see, for example, Andronov and Witt 1930; Arnold 1974; Rabinovich and Trubetskov 1989, and references therein). Synchronous periodic states have been observed in many dynamical processes in various fields of science and engineering, and synchronization of coupled periodic oscillators is now understood as the ability of the oscillators to switch their behaviors from quasiperiodicity (motion on a stable torus) to limit cycles (frequency locking), as the coupling strength is increased. For a comprehensive review on this topic, we refer a reader to the excellent monograph of Pikovsky et al. (2001).

Here, we mainly focus on synchronization of chaotic systems. Simple systems of two coupled chaotic oscillators can already show many types of synchronization, while some other synchronous states emerge only in high-dimensional systems formed by three (or more) oscillators coupled in a network.

This chapter starts with considerations about synchronization of periodic phase oscillators; then, we continue with the description of different types of synchronization of chaotic systems. These are complete synchronization (Pecora and Carroll 1990), phase synchronization (Rosenblum et al. 1996), antiphase synchronization (Liu et al. 2006), lag synchronization (Rosenblum et al. 1997), anticipating synchronization (Voss 2000), and generalized synchronization (Rulkov et al. 1995).

In addition to the steady synchronous states mentioned above, one can also find unstable synchronous states, such as in intermittent synchronization (Gauthier and Bienfang 1996) and episodic synchronization (Buldú et al. 2006). These happen when any kind of synchronization is interrupted by asynchronous oscillations or the system intermittently changes synchronization type, for instance switching from phase synchronization to lag synchronization and back (Pisarchik and Jaimes-Reátegui 2005b).

2.2 Types of Coupling

Let us start by introducing a definition of coupling.

An N-dimensional dynamical system is called *uncoupled* if it can be decomposed in $k < N$ independent dynamical subsystems with total dimension equal to N. In the simplest case of two oscillators ($k = 2$), an uncoupled system can be presented in the form

$$\begin{aligned} \dot{\mathbf{x}} &= \mathbf{F}(\mathbf{x}), \\ \dot{\mathbf{y}} &= \mathbf{G}(\mathbf{y}), \end{aligned} \tag{2.1}$$

where $\mathbf{x} \in \mathbb{R}^n$, $\mathbf{y} \in \mathbb{R}^m$, and the total dimension of the coupled system is $N = n+m$.

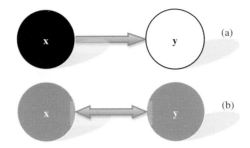

Figure 2.2 The two possible configurations for two coupled oscillators. (a) Unidirectional coupling, or master–slave configuration. (b) Bidirectional coupling.
In the former case, the master and slave oscillators are shown, respectively, in
black and white. In the latter, the two oscillators, which mutually combine their
influence and interaction, are shown in gray.

If we work with two oscillators, there are only two possible coupling configurations: the *unidirectional* (Figure 2.2(a)) and the *bidirectional* (Figure 2.2(b))
coupling schemes.

A *unidirectionally coupled* dynamical system is described by

$$\dot{\mathbf{x}} = \mathbf{F}(\mathbf{x}),$$
$$\dot{\mathbf{y}} = \mathbf{G}(\mathbf{y}) + \mathbf{K}(\mathbf{x}, \mathbf{y}),$$
(2.2)

where $\mathbf{K}(\mathbf{x}, \mathbf{y})$ is a nonzero function of $\mathbf{x}$ and $\mathbf{y}$. This coupling scheme is often
referred to as *master–slave configuration*, where the subsystem $\mathbf{x}$ is called *master*
or *drive* and the subsystem $\mathbf{y}$ is called *slave* or *response*. In some region of the phase
space $\mathbb{R}^{n+m}$, the behavior of the slave is influenced by the behavior of the master,
while the driving system is completely independent of the response system.

A *bidirectionally coupled* dynamical system is described by

$$\dot{\mathbf{x}} = \mathbf{F}(\mathbf{x}) + \mathbf{K}_1(\mathbf{y}, \mathbf{x}),$$
$$\dot{\mathbf{y}} = \mathbf{G}(\mathbf{y}) + \mathbf{K}_2(\mathbf{x}, \mathbf{y}),$$
(2.3)

where $\mathbf{K}_1(\mathbf{y}, \mathbf{x})$ and $\mathbf{K}_2(\mathbf{x}, \mathbf{y})$ are nonzero functions of $\mathbf{x}$ and $\mathbf{y}$. If the coupling is
symmetric, then $\mathbf{K}_1 = \mathbf{K}_2$. In this case, each subsystem influences the other in
some region of the phase space $\mathbb{R}^{n+m}$.

The coupling function can be defined in many different ways. The most popular way to couple two subsystems is the so-called *diffusive coupling*, which
corresponds to setting

$$\mathbf{K}(\mathbf{x}, \mathbf{y}) = \sigma(\mathbf{H}(\mathbf{x}) - \mathbf{y}),$$

where $0 \leqslant \sigma \leqslant 1$ is the coupling strength and $\mathbf{H}$ is a response function. The scalar
constant σ quantifies the coupling strength: when uncoupled ($\sigma = 0$) the systems

follow their own dynamics (Crank 1980). Note that in real-world systems the diffusion is usually asymmetric, $K_1 \neq K_2$ (Bragard et al. 2007). A typical example is the case of physiological membranes that selectively diffuse ions (Nelson 2003).

When three or more oscillators are accounted for, a larger number of coupling configurations can be realized. In the theory of graphs or complex networks, these are called *network motifs*. A motif is a pattern of interconnections (links) which occurs significantly more often than in randomized versions of the same graph, i.e., in graphs with the same number of nodes, links and degree distribution as the original one, but where the links are randomly distributed.

This notion of network motifs was used by Uri Alon and his colleagues (Shen-Orr et al. 2002) to characterize patterns of interconnections in the gene regulation (transcription) network of the bacterium *Escherichia coli*, establishing an important link between mathematics and biology. Later, other researchers focused on computational applications of network motifs, applying their study to the analysis of the internal structure of computer software (Valverde and Solé 2005).

There are 13 possible motifs formed by three coupled oscillators, as shown in Figures 2.3 and 2.4. They can be coupled either in a line (Figure 2.3) or in a ring (Figure 2.4). The inline coupled oscillators can form six possible motifs: *relay* (Figure 2.3(a)), *chain* (Figure 2.3(b)), *competitive* (Figure 2.3(c)), *auxiliary* (Figure 2.3(d)), and two *mixed* (Figure 2.3(e, f)) coupling schemes. The mixed mode is twice degenerate because it is formed by two distinct motifs: one of the oscillators can be either master only (Figure 2.3(e)) or slave only (Figure 2.3(f)). The ring coupling, instead, can be mutual (all-to-all), unidirectional, or combined (Figure 2.4). Note that the combined coupling can be achieved by four distinct motifs, and that in counting the motifs, the graph is considered unlabeled. In addition, feedback loops with time-delayed coupling can be added onto each of the coupled subsystems.

The consideration of these simple schemes is very important for further understanding the mechanisms leading to synchronization of complex networks formed by many coupled oscillators, discussed in detail in Chapter 5, where such simple systems are elementary cells.

We start here by describing the case of periodic systems, and then continue with chaotic dynamics. To provide examples for the main concepts, we refer to the Kuramoto (1975) and Rössler (1976) models, since they are paradigms for a phase and a chaotic oscillator, respectively. Later on, we will discuss different types of synchronization in more sophisticated systems used to model physical and biological processes, and present experimental evidence for some kinds of synchronous behavior.

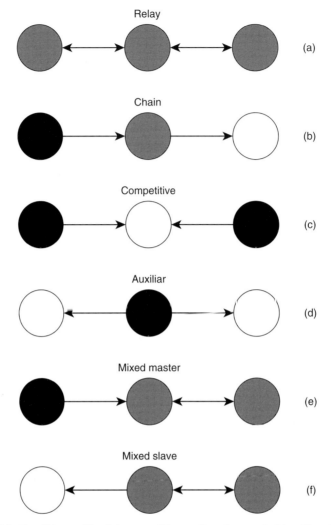

Figure 2.3 Possible motifs of three oscillators directly coupled in a line. (a) Relay,
(b) chain, (c) competitive, (d) auxiliar, and (e, f) mixed types of coupling schemes.
The color code is as follows: black for master systems, white for slave systems,
and grey for systems that are mutually interacting.

2.3 Phase Oscillators

One of the simplest types of synchronization is *frequency entrainment*, or *frequency
locking*, defined as the ability of oscillators to adjust their behaviors to a common
frequency as the coupling strength is increased. This generic type of synchronization occurs in both periodic and chaotic oscillators. Furthermore, frequency
entrainment can also be achieved in oscillators driven by a periodic force applied
to a particular system parameter or variable.

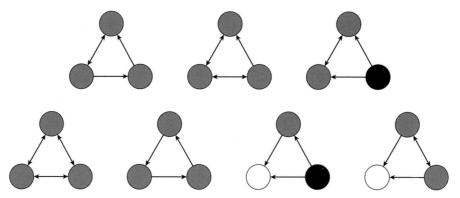

Figure 2.4 Motifs formed by three oscillators coupled in a ring. The color code is: black for master systems, white for slave systems, and grey for systems that are mutually interacting.

In order to understand the simplest manifestations of these phenomena, we introduce the concept of a *phase oscillator*, i.e., a dynamical system with only one periodic dynamical variable ϑ, which is called the phase of the oscillator.

A single phase oscillator fulfills the dynamical equation

$$\dot{\vartheta} = F(\vartheta, t), \tag{2.4}$$

where $F(\vartheta, t)$ is a real function that is 2π-periodic in ϑ. A trivial example is the *uniform* phase oscillator $F(\vartheta, t) = \omega$, where the angular frequency ω is a real constant. The period of one oscillation is in this case $T = 2\pi/\omega$.

Coupled ensembles of phase oscillators are particularly suited to describe the emergence of synchronous behavior in a system of many interacting units. The paradigmatic model in this case is the *Kuramoto model* (Kuramoto 1975), described by

$$\dot{\vartheta}_i = \omega_i + \frac{K}{N} \sum_{j=1}^{N} \sin(\vartheta_j - \vartheta_i) \quad i = 1, \dots, N, \tag{2.5}$$

where ϑ_i is the phase of the ith oscillator with natural frequency ω_i, and K is the coupling strength.

This model assumes an all-to-all interaction among the elements of an ensemble of oscillators. The interaction between each pair of oscillators depends sinusoidally on the difference between their phases. The Kuramoto model has found widespread applications, especially to describe collective behavior in chemical and biological systems.

While a deeper discussion on the Kuramoto model will be presented in Chapter 4, here we limit ourselves to study a number of simplified variants of Equation 2.5 for the simple and fully solvable case of $N = 2$.

2.3.1 Unidirectionally Coupled Phase Oscillators

As a first example, consider a system of two unidirectionally coupled phase oscillators (Figure 2.2(a))

$$
\begin{aligned}
\dot{\vartheta}_1 &= \omega_1, \\
\dot{\vartheta}_2 &= \omega_2 + \sigma \sin(\vartheta_1 - \vartheta_2).
\end{aligned}
\tag{2.6}
$$

where the subscripts 1 and 2 refer to the master and slave oscillator, respectively. There are two frequency parameters ω_1 and ω_2 and a coupling parameter σ. The coupling term is designed to be similar to the Kuramoto model of Equation 2.5.

The phases ϑ_1 and ϑ_2 are synchronized if their difference $\varphi(t) = \vartheta_2(t) - \vartheta_1(t)$ is bounded for $t > 0$. The phase oscillator ϑ_1 is uniform and therefore has the solution $\vartheta_1(t) = \omega_1 t + \vartheta_1(0)$, where $\vartheta_1(0)$ is its initial phase at time 0.

The coupling term appearing in the equation for $\dot{\vartheta}_2$ is designed in such a way that it only depends on the difference $\vartheta_1 - \vartheta_2$ of the phase variables. This allows one to rewrite the system in terms of a phase difference variable $\varphi = \vartheta_2 - \vartheta_1$, and obtain the *Adler equation* (Adler, 1946)[1]

$$
\dot{\varphi} = \Delta - \sigma \sin \varphi.
\tag{2.7}
$$

The parameter $\Delta = \omega_2 - \omega_1$ is called the *detuning* between the two phase oscillators.

The right-hand side of Equation 2.7 is shown in Figure 2.5 for a number of different parameter values. We see that for strong coupling ($\sigma/|\Delta| > 1$, right

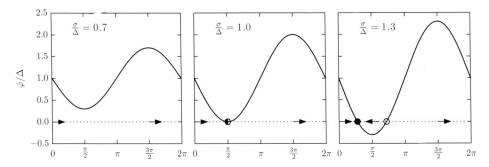

Figure 2.5 $\dot{\varphi}/\Delta$ of the Adler Equation 2.7 for a number of different values of σ/Δ. The stability of the fixed points is indicated by $\bigcirc$ (unstable), $\bullet$ (stable) and $\circleddash$ (half stable from left).

---

[1] In 1946, the American inventor of Austrian origin, Robert Adler, derived a nonlinear differential equation for the oscillators' phase difference with an injection signal to describe the phase-locking phenomenon in a cross-coupled LC oscillator. He also recognized that this equation could model a pendulum suspended in a viscous fluid inside a rotating container.

panel), there are two fixed points, one stable $\varphi_0^s = \arcsin(\Delta/\sigma)$ and one unstable $\varphi_0^u = \pi - \arcsin(\Delta/\sigma)$. Therefore, the phase difference φ asymptotically approaches the stable fixed point φ_0^s for all initial conditions except φ_0^u. As a consequence, all solutions of Equations 2.6 approach $\vartheta_2(t) = \vartheta_1(t) + \varphi_0^s$, i.e., both phases ϑ_1 and ϑ_2 eventually rotate with the same frequency ω_1. This regime is known as *frequency entrainment* or *phase-locking* regime, because the phase difference φ is independent of time.

As the coupling is decreased, there exists a critical value

$$\sigma_c = |\Delta| \qquad (2.8)$$

with only one fixed point, which is at $\varphi_0^{hs} = \pi/2$ for positive detuning and at $\varphi_0^{hs} = 3\pi/2$ for negative detuning. This fixed point is half-stable, as illustrated in the middle panel of Figure 2.5.

At σ_c, a saddle-node bifurcation occurs and therefore (for $|\sigma| < \sigma_c$, left panel of Figure 2.5) the stable and unstable fixed points annihilate each other, and there exist no other fixed points for φ. In this regime, the phase difference $|\varphi(t)|$ grows without bounds (modulo 2π), and one oscillator repeatedly passes the other. This phenomenon is called *unlocking* or *unbounded phase oscillations*.

To better understand this case, let us assume that $\Delta > \sigma > 0$, and furthermore that at $t = 0$ both oscillators ϑ_1 and ϑ_2 start out with the same phase. Since $\dot\varphi > 0$, the second oscillator advances more quickly than the first. However, it is important to note that $\dot\varphi$ is not constant. In fact, from Equation 2.7 and from the left panel of Figure 2.5, it is easy to see that $\dot\varphi$ has a minimum value of $\Delta - \sigma$ at $\varphi = \pi/2$. If Δ is only slightly larger than σ, then the system will spend a long time close to this minimum. In this case we speak of a "slow region" at $\varphi = \pi/2$. In a less formal language, we can say that the "ghost" of the deceased half-stable point φ_0^{hs} that existed at the bifurcation point σ_c continues to slow down the phase difference variable φ, even though it does not have the power to stop it.

Only when φ reaches the value of 2π, the phases of both oscillators will again coincide. We then say that a *phase slip* has occurred. The required time T_φ for φ to increase from 0 to 2π is given by

$$T_\varphi = \int_0^{2\pi} \frac{d\varphi}{\Delta - \sigma \sin \varphi}. \qquad (2.9)$$

To compute this integral, let $x = \tan(\varphi/2)$. Then, it is

$$\frac{d}{d\varphi}x = \frac{1}{2}\left(1 + \tan^2 \frac{\varphi}{2}\right) = \frac{1}{2}\left(1 + x^2\right),$$

hence

$$d\varphi = \frac{2}{1 + x^2}dx.$$

Also, since $\sin \varphi = 2 \sin (\varphi/2) \cos (\varphi/2)$, it is

$$\sin \varphi = 2x \cos^2 \frac{\varphi}{2} = \frac{2x}{1 + x^2},$$

where we used the identity $\cos^2 \alpha = 1 + \tan^2 \alpha$. Then, we obtain

$$
\begin{aligned}
T_\varphi &= \int_{-\infty}^{\infty} \left(\Delta - \sigma \frac{2x}{1 + x^2} \right)^{-1} \frac{2}{1 + x^2} dx \\
&= \int_{-\infty}^{\infty} \left(\frac{\Delta + \Delta x^2 - 2\sigma x}{1 + x^2} \right)^{-1} \frac{2}{1 + x^2} dx \\
&= 2 \int_{-\infty}^{\infty} \frac{1}{\Delta + \Delta x^2 - 2\sigma x} dx \\
&= \frac{2}{\Delta} \int_{-\infty}^{\infty} \frac{1}{x^2 - \frac{2\sigma}{\Delta} x + 1} dx \\
&= \frac{2}{\Delta} \int_{-\infty}^{\infty} \frac{1}{\left(x^2 - \frac{2\sigma}{\Delta} x + \frac{\sigma^2}{\Delta^2} \right) - \frac{\sigma^2}{\Delta^2} + 1} dx \\
&= \frac{2}{\Delta} \int_{-\infty}^{\infty} \frac{1}{\left(x + \frac{\sigma}{\Delta} \right)^2 + \frac{\Delta^2 - \sigma^2}{\Delta^2}} dx.
\end{aligned}
$$

Now let $u = x + \sigma/\Delta$ and $v^2 = \left(\Delta^2 - \sigma^2 \right) / \Delta^2$ to get

$$
\begin{aligned}
T_\varphi &= \frac{2}{\Delta} \int_{-\infty}^{\infty} \frac{1}{u^2 + v^2} du \\
&= \frac{2}{\Delta} \frac{1}{v} \arctan \frac{u}{v} \Big|_{-\infty}^{\infty} \\
&= \frac{2}{\Delta} \frac{\Delta}{\sqrt{\Delta^2 - \sigma^2}} \left(\frac{\pi}{2} + \frac{\pi}{2} \right) \\
&= \frac{2\pi}{\sqrt{\Delta^2 - \sigma^2}}.
\end{aligned}
$$

This means that during the time interval T_φ the phase variables ϑ_1 and ϑ_2 have increased by $\omega_1 T_\varphi$ and $\omega_1 T_\varphi + 2\pi$, respectively. Their average phase velocities are therefore given by

$$\langle \Omega_1 \rangle \equiv \langle \dot{\vartheta}_1 \rangle = \omega_1, \tag{2.10}$$

$$\langle \Omega_2 \rangle \equiv \langle \dot{\vartheta}_2 \rangle = \frac{\omega_1 T_\varphi + 2\pi}{T_\varphi} = \omega_1 + \Delta \sqrt{1 - \frac{\sigma^2}{\Delta^2}}. \tag{2.11}$$

Equation 2.11 also holds for negative Δ.

2.3.2 Mutually Coupled Phase Oscillators

Let us now consider the case of two mutually coupled phase oscillators as in Figure 2.2(b) and study the set of equations

$$\dot{\vartheta}_1 = \omega_1 + \sigma \sin(\vartheta_2 - \vartheta_1),$$
$$\dot{\vartheta}_2 = \omega_2 + \sigma \sin(\vartheta_1 - \vartheta_2). \tag{2.12}$$

This is a special case of the Kuramoto model (Kuramoto, 1975), which we will discuss in more detail in Chapter 4. When uncoupled ($\sigma = 0$), the phases rotate at natural frequencies ω_1 and ω_2. To solve the case with coupling, we introduce the phase difference variable $\varphi = \vartheta_2 - \vartheta_1$ and a phase average variable $\Phi(t) = (\vartheta_2 + \vartheta_1)/2$ and obtain the set of equations

$$\dot{\varphi} = \Delta - 2\sigma \sin \varphi, \tag{2.13}$$
$$\dot{\Phi} = \Omega, \tag{2.14}$$

with $\Delta = \omega_2 - \omega_1$ and $\Omega = (\omega_2 + \omega_1)/2$.

The dynamical variables φ and Φ are uncoupled. Φ is a uniform phase oscillator, and the comparison of Equation 2.13 with Adler Equation 2.7 shows that they have the same form, except for a factor of 2 in front of the coupling term.

In analogy with the results for the unidirectionally coupled system, the phase variables ϑ_1 and ϑ_2 asymptotically approach a phase locked state $\vartheta_2(t) = \vartheta_1(t) + \arcsin(\Delta/2\sigma)$ if $\sigma > |\Delta|/2$. In this case, both oscillators share the common frequency Ω.

Conversely, following the same procedure as in the unidirectional coupling case, one finds that, for low coupling $\sigma < |\Delta|/2 = \sigma_c$, the average frequencies are

$$\langle \Omega_{1,2} \rangle \equiv \langle \dot{\vartheta}_{1,2} \rangle = \Omega \pm \frac{\Delta}{2} \sqrt{1 - \frac{4\sigma^2}{\Delta^2}}. \tag{2.15}$$

2.3.3 Frequency-Splitting Bifurcation

We have seen that two phase oscillators, either unidirectionally (Equation 2.6) or mutually (Equation 2.12) coupled, undergo a desynchronization transition at $\sigma = \sigma_c$, which is caused by a saddle-node bifurcation in the phase difference variable φ.

After the fixed points have disappeared, the two oscillators "split" their average frequencies according to Equations 2.11 and 2.15. This transition is therefore also called a *frequency-splitting bifurcation* (Maistrenko et al. 2005). Figure 2.6 shows the two typical bifurcation diagrams of the average frequencies (when the coupling strength is taken as a control parameter) for unidirectional (Figure 2.6(a)) and bidirectional (Figure 2.6(b)) coupling.

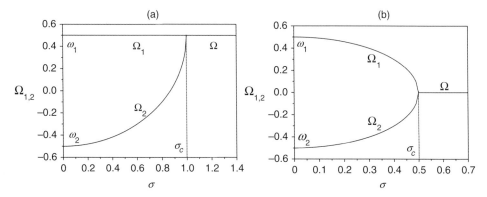

Figure 2.6 Frequency-splitting bifurcation diagrams of rotation frequencies for coupled phase oscillators using the coupling strength σ as a control parameter for (a) unidirectional coupling Equation 2.6 and (b) bidirectional coupling Equation 2.12. The natural frequencies are $\omega_1 = 0.5$ and $\omega_2 = -0.5$.

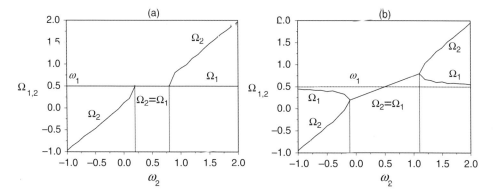

Figure 2.7 Frequency-splitting bifurcation diagrams obtained using the natural frequency ω_2 as a control parameter (while $\omega_1 = 0.5$ is kept fixed) for (a) unidirectional coupling and (b) bidirectional coupling. The coupling strength is fixed at $\sigma = 1.4$. The two bifurcations marked by the vertical dotted lines bound the region of frequency entrainment where $\Omega_2 = \Omega_1$.

The transition from asynchronous to synchronous behavior is also observed when the natural frequency of one of the two oscillators (say, ω_1) is fixed, while the natural frequency of the other oscillator (ω_2) is varied.

The corresponding bifurcation diagrams of the average rotation frequencies Ω_1 and Ω_2 for unidirectional (Figure 2.7(a)) and bidirectional (Figure 2.7(b)) coupling display two frequency-splitting bifurcations that bound the region of frequency entrainment $|\Delta| < \sigma$. Note that the region of frequency entrainment in the case of bidirectional coupling is double the range of that for unidirectional coupling, because in the former case the coupling is applied to both oscillators (compare Equations 2.6 and 2.12).

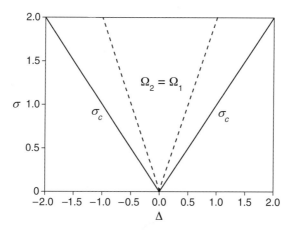

Figure 2.8 Frequency-locking range in the space of frequency detuning Δ and coupling strength σ for unidirectional (dashed lines) and bidirectional (solid lines) coupling. The Arnold tongue is the triangular region bounded by the lines σ_c, inside which the phases rotate with the same frequency $\Omega_2 = \Omega_1$.

The phases of the two oscillators coincide ($\varphi = 0$) if and only if $\Delta = 0$, i.e., when the oscillators are identical. In this case we deal with *identical* or *complete synchronization*, which is only possible for completely identical oscillators.

In the case of periodic oscillators, frequency-locking is always accompanied by *phase-locking*, meaning that the phase difference is constant in time. The phase-locking range becomes larger for stronger coupling, forming a tongue in the (Δ, σ) parameter space, as shown in Figure 2.8. This tongue is named *Arnold tongue* in honor of Russian mathematician Vladimir Arnold (1937–2010), who discovered phase-locking behavior in periodically forced systems.

Some other interesting synchronization states occur in the case of unidirectional coupling (for nonzero initial detuning) inside the frequency-locking range. If $\omega_1 < \omega_2$, the phase difference is negative, otherwise it is positive. In the former case ($\varphi < 0$) we deal with so-called *lag synchronization*, whereas the latter case ($\varphi > 0$) is referred to as *anticipating synchronization*. In both cases, the state of one oscillator anticipates the state of the other. For lag synchronization, the master anticipates the slave, whereas for anticipating synchronization the slave system anticipates the state of the master.

At this stage, it is worth noticing that, through the use of such a simple example as only two coupled phase oscillators, five different types of synchronization can already be described: frequency entrainment, phase-locking, lag, anticipating, and identical (complete) synchronization. In addition, although frequency entrainment and phase-locking are equivalent for periodic oscillators, they may exhibit completely different behaviors in chaotic oscillators.

2.4 Complete Synchronization

The notion of synchronization has been extended to chaotic dynamics since the appearance of the work of Fujisaka and Yamada (1983), who first demonstrated that two identical chaotic systems can change their individual behaviors from uncorrelated oscillations to completely identical oscillations as the coupling strength is increased. In the same period, Afraimovich et al. (1986) have defined synchronization of chaotic systems in the sense of attractor dimension, considering two coupled chaotic systems to be synchronized if the dimension of the combined attractor equals the dimensions of its projections into the subspaces of the coupled subsystems.

Here, we begin our discussion of synchronization in chaotic systems with the strongest synchronization type emerging in systems of identical coupled elements, namely *complete* or *identical synchronization*. This type of synchronization was first analyzed by Kaneko (1986) and Afraimovich et al. (1986) in both discrete and continuous systems. Its defining feature is that, starting from different initial conditions, the state variables of identical chaotic systems coincide as time tends to infinity.

In other words, complete synchronization implies that the two chaotic systems eventually evolve identically in time, despite their characteristic dependence on the initial conditions. Complete synchronization has been widely used for different applications, such as secure communication, chaos suppression, and monitoring chaotic dynamics (Hayes et al. 1993; Argyris et al. 2005).

The general case can be described by considering two coupled nonlinear systems, given by

$$\dot{\mathbf{x}} = \mathbf{g}(\mathbf{x}) + \hat{\mathbf{C}}_{\mathbf{x}}\mathbf{v}(\mathbf{x}, \mathbf{y}), \tag{2.16}$$

and

$$\dot{\mathbf{y}} = \mathbf{h}(\mathbf{y}) + \hat{\mathbf{C}}_{\mathbf{y}}\mathbf{u}(\mathbf{x}, \mathbf{y}), \tag{2.17}$$

where $\mathbf{x} \in \mathbb{R}^n$ and $\mathbf{y} \in \mathbb{R}^m$ are vector variables, $\mathbf{g}$ and $\mathbf{h}$ are vector functions of the system variables, $\mathbf{v}$ and $\mathbf{u}$ are coupling functions, and $\hat{\mathbf{C}}_{\mathbf{x}}$ and $\hat{\mathbf{C}}_{\mathbf{y}}$ are matrices of coupling coefficients of appropriate dimension. For symmetric coupling we have $\mathbf{v} = -\mathbf{u}$ and $\hat{\mathbf{C}}_{\mathbf{x}} = \hat{\mathbf{C}}_{\mathbf{y}} \equiv \hat{\mathbf{C}}$. Unidirectional coupling implies $\hat{\mathbf{C}}_{\mathbf{x}} = 0$.

2.4.1 Measures of Complete Synchronization

The coupled system of Equations 2.16 and 2.17 is said to be completely synchronized if all their state variables asymptotically match, i.e.,

$$\lim_{t \to \infty} |\mathbf{x}(t) - \mathbf{y}(t)| = 0. \tag{2.18}$$

Rigorously, this state is only possible for identical systems ($\mathbf{g} = \mathbf{h} \equiv \mathbf{f}$).

The simplest way to detect complete synchronization is the direct comparison of the state variables of the interacting systems **x** and **y**, by calculating the *synchronization error* defined as

$$\mathbf{e}(t) = \mathbf{y}(t) - \mathbf{x}(t), \tag{2.19}$$

and monitoring its temporal evolution.

Complete synchronization can be also estimated using the *cross-correlation* between the vector variables **x**(*t*) and **y**(*t*), given as

$$\mathbf{C}(\tau) = \int_{-\infty}^{\infty} \mathbf{x}^*(t)\mathbf{y}(t + \tau)\, \mathrm{d}t, \tag{2.20}$$

where * denotes the complex conjugate and τ is a time shift between the two signals.

In practical situations when only one variable per system is accessible, the cross-correlation can be computed from scalar time series of the available variables $x(t)$ and $y(t)$ of the coupled systems Equations 2.16 and 2.17, as

$$C(\tau) = \frac{\langle [x(t) - \langle x(t)\rangle][y(t + \tau) - \langle y(t)\rangle]\rangle}{\sqrt{\langle [x(t) - \langle x(t)\rangle]^2\rangle\langle [y(t) - \langle y(t)\rangle]^2\rangle}}, \tag{2.21}$$

where the angled brackets denote time average.

Both quantities are useful to estimate synchronization quality. Complete synchronization is achieved when $\mathbf{e}(t) \to \mathbf{0}$ or $\mathbf{C}(\tau) \to 1$ for $t \to \infty$.

2.4.2 Bidirectional Coupling

Bidirectional coupling is possibly the easier way to achieve complete synchronization. According to Fujisaka and Yamada (1983), a way to introduce bidirectional coupling between two identical chaotic systems is by adding symmetric linear coupling terms to the expressions that define their evolution. This type of coupling mechanism, which may be total or partial, is known as *linear diffusive coupling*.

A bidirectional symmetric coupling scheme between identical chaotic systems is tantamount to introducing additional dissipation in the dynamics:

$$\begin{aligned}\dot{\mathbf{x}} &= \mathbf{f}(\mathbf{x}) + \hat{\mathbf{C}}(\mathbf{y} - \mathbf{x}), \\ \dot{\mathbf{y}} &= \mathbf{f}(\mathbf{y}) + \hat{\mathbf{C}}(\mathbf{x} - \mathbf{y}),\end{aligned} \tag{2.22}$$

where $\mathbf{x} \in \mathbb{R}^n$, $\mathbf{y} \in \mathbb{R}^n$ and $\hat{\mathbf{C}}$ is a $n \times n$ diffusion matrix whose elements determine the dissipative coupling.

When increasing the coupling strength (given by the elements of $\hat{\mathbf{C}}$), the system in Equation 2.22 displays a transition to a completely synchronized state $\mathbf{x}(t) = \mathbf{y}(t) = \mathbf{X}(t)$, and the motion takes place on an invariant subspace of the coupled system referred to as *synchronization manifold*, described by a system $\dot{\mathbf{X}} = \mathbf{F}(\mathbf{X})$. If this subspace is locally attractive, the coupled system exhibits identical synchronization. Furthermore, if the synchronization occurs independently of initial conditions, it is referred to as *global* synchronization.

To study the stability of the synchronization manifold, we have to analyze the motion of the coupled system in Equation 2.22 after a small perturbation:

$$\begin{aligned}
\mathbf{x}(t) &= \mathbf{X}(t) + \delta\mathbf{x}(t), \\
\mathbf{y}(t) &= \mathbf{X}(t) + \delta\mathbf{y}(t),
\end{aligned} \tag{2.23}$$

where $\delta\mathbf{x}(t)$ and $\delta\mathbf{y}(t)$ are the projections of the small perturbation onto the $\mathbf{x}$ and $\mathbf{y}$ subspaces. The synchronization manifold $\mathbf{X}(t)$ is locally stable if any small perturbation brings the system back to the synchronous state $\mathbf{X}(t)$, i.e., $\mathbf{X}(t)$ is an attractor.

If the perturbation is small enough, one can use a linear approximation and only consider the first-order terms in the time derivatives of the perturbation projections:

$$\begin{aligned}
\delta\dot{\mathbf{x}} &= \mathbf{J}_{\mathbf{F}}(\mathbf{X})\delta\mathbf{x} + \hat{\mathbf{C}}(\delta\mathbf{y} - \delta\mathbf{x}), \\
\delta\dot{\mathbf{y}} &= \mathbf{J}_{\mathbf{F}}(\mathbf{X})\delta\mathbf{y} + \hat{\mathbf{C}}(\delta\mathbf{x} - \delta\mathbf{y}),
\end{aligned} \tag{2.24}$$

where $\mathbf{J}_{\mathbf{F}}$ is the Jacobian matrix at the trajectory $\mathbf{X}(t)$.

Instead of projecting the perturbation onto the subspaces of the variables $\mathbf{x}$ and $\mathbf{y}$, it is useful to consider the system evolution with respect to the perturbation projections onto the synchronization manifold and the transverse subspace, given as

$$\begin{aligned}
\delta\mathbf{X}_{\parallel} &= (\delta\mathbf{x} + \delta\mathbf{y})/\sqrt{2}, \\
\delta\mathbf{X}_{\perp} &= (\delta\mathbf{x} - \delta\mathbf{y})/\sqrt{2},
\end{aligned} \tag{2.25}$$

and called, respectively, parallel ($\delta\mathbf{X}_{\parallel}$) and transverse ($\delta\mathbf{X}_{\perp}$) perturbations. The time evolution of these perturbations can be written as

$$\begin{aligned}
\delta\dot{\mathbf{X}}_{\parallel} &= \mathbf{J}_{\mathbf{F}}(\mathbf{X}) \cdot \delta\mathbf{X}_{\parallel}, \\
\delta\dot{\mathbf{X}}_{\perp} &= [\mathbf{J}_{\mathbf{F}}(\mathbf{X}) - 2\hat{\mathbf{C}}] \cdot \delta\mathbf{X}_{\perp}.
\end{aligned} \tag{2.26}$$

The system is stable if the perturbations perpendicular to the synchronization manifold exponentially vanish, i.e., if all Lyapunov exponents for the transversal part of Equation 2.26 are negative.

2.4.3 Unidirectional Coupling

The unidirectional coupling case, also called *master–slave* or *drive-response configuration*, is also interesting, due to its promising applications in secure communication.

In the master–slave configuration, the behavior of the response system is dependent on the behavior of the drive, but the latter does not influence the former. The stability of the response system can be characterized by the *conditional Lyapunov exponents* introduced by Pecora and Carroll (1990), and derived from variational equations of the response system. The term "conditional" is used because these exponents depend on the variables of the response system only and are independent of those of the drive system. Sometimes they are also called *transversal Lyapunov exponents* because they correspond to directions transverse to the synchronization manifold (Güemez and Matías 1995; Gauthier and Bienfang 1996). As mentioned above, synchronization is achieved if all conditional Lyapunov exponents are negative.

Among different unidirectional coupling types, we will consider the most commonly used schemes. These are the *homogeneous driving configuration* (Pecora and Carroll 1991; Cuomo and Oppenheim 1993), the *active–passive decomposition* (Kocarev and Parlitz 1995; Parlitz et al. 1996a), and the *Fujisaka–Yamada scheme* (Fujisaka and Yamada 1983; Pyragas 1993; Brown et al. 1994).

The first method is based on the concept of a response system locking onto a driver system (Pecora and Carroll 1990; He and Vaidya 1992), i.e., the ith variable of the response system is simply replaced by the corresponding variable of the drive system. The second method consists in a decomposition of a given chaotic system into active and passive parts. Finally, the last method involves the application of a control signal proportional to the difference between the drive and response variables. Let us consider the main principles of these methods.

Homogeneous Driving Configuration

Pecora and Carroll (1990) proposed the construction of a chaotic synchronizing system by decomposing an autonomous n-dimensional dynamical system

$$\dot{\mathbf{u}} = \mathbf{F}(\mathbf{u}), \tag{2.27}$$

into drive

$$\dot{\mathbf{x}} = \mathbf{V}(\mathbf{x}), \qquad \mathbf{x} \in \mathbb{R}^m \tag{2.28}$$

and response

$$\dot{\mathbf{y}} = \mathbf{H}(\mathbf{y}, \mathbf{x}), \ \mathbf{y} \in \mathbb{R}^k \tag{2.29}$$

subsystems, with $m + k = n$. Here, $\mathbf{u} \equiv (u_1, \ldots, u_n)$ is an n-dimensional state vector, $\mathbf{F}$ is a vector function from $\mathbb{R}^n$ to $\mathbb{R}^n$, $\mathbf{x} \equiv (x_1, \ldots, x_m) = (u_1, \ldots, u_m)$ and $\mathbf{y} \equiv (y_1, \ldots, y_k) = (u_{m+1}, \ldots, u_n)$ are the drive and response state vectors, $\mathbf{V} \equiv (F_1(\mathbf{x}), \ldots, F_m(\mathbf{x}))$, and $\mathbf{H} \equiv (F_{m+1}(\mathbf{u}), \ldots, F_n(\mathbf{u}))$.

This scheme is particularly useful when the subsystems are identical and coupled in only one variable x_j. In this case, $m = k$, $\mathbf{V} = \mathbf{H}$, and x_j can be considered as an external chaotic driving. This allows one to create a new subsystem with dimension $m + 1$:

$$\dot{\mathbf{y}}' = \mathbf{h}(\mathbf{y}, x_j), \tag{2.30}$$

and consider its synchronization with the response subsystem of Equation 2.29.

In this context, complete synchronization is defined as the identity between the trajectories of the response system $\mathbf{y}$ and its replica $\mathbf{y}'$ for the same chaotic driving signal x_j. These subsystems synchronize if $|\mathbf{e}(t)| \equiv |\mathbf{y} - \mathbf{y}'| \to 0$. This occurs if and only if the conditional Lyapunov exponents of the response system $\mathbf{y}$ are all negative. Notice that not all possible selections of the driving signal lead to a synchronized state.

Active–Passive Decomposition

An alternative coupling configuration for the drive-response scheme was proposed by Kocarev and Parlitz (1995). The method involves the decomposition of a chaotic autonomous system

$$\dot{\mathbf{z}} = \mathbf{F}(\mathbf{z}) \tag{2.31}$$

into two identical nonautonomous subsystems

$$\dot{\mathbf{x}} = \mathbf{f}(\mathbf{x}), \mathbf{s}(t)), \tag{2.32}$$
$$\dot{\mathbf{y}} = \mathbf{f}(\mathbf{y}), \mathbf{s}(t)), \tag{2.33}$$

driven by the same force $\mathbf{s}(t) = \mathbf{h}(\mathbf{x})$.

The systems described by Equations 2.32 and 2.33 synchronize if the error approaches zero ($\mathbf{e} = \mathbf{x} - \mathbf{y} \to \mathbf{0}$) when $t \to \infty$, i.e., if the differential equation

$$\dot{\mathbf{e}} = \mathbf{f}(\mathbf{x}, \mathbf{s}) - \mathbf{f}(\mathbf{y}, \mathbf{s}) = \mathbf{f}(\mathbf{x}, \mathbf{s}) - \mathbf{f}(\mathbf{x} - \mathbf{e}, \mathbf{s}) \tag{2.34}$$

possesses a stable fixed point at $\mathbf{e} = 0$. In some cases this can be proven using linear stability analysis of Equation 2.34. In general, the stability of the synchronization manifold $\mathbf{x} = \mathbf{y}$ can be found numerically. Synchronization occurs if all conditional Lyapunov exponents of the nonautonomous system Equation 2.32 are negative under the explicit constrain that they must be calculated on the trajectory $\mathbf{s}(t)$. In this sense, the solution of the system Equation 2.32 tends to a fixed point when there is no driving.

Fujisaka–Yamada Scheme

Many researchers use the coupling scheme proposed by Fujisaka and Yamada (1983) and later developed by other scientists (Pyragas 1993; Brown et al. 1994).

In such a scheme, the coupling is proportional to the difference between some variables of the coupled oscillators **u** and **v**.

This method can be simply achieved by defining a coupling constant σ and adding the scalar function $\sigma(u_j(t) - v_j(t))$ to the jth component of the vector field $\dot{\mathbf{u}} = \mathbf{G}(\mathbf{u})$ of the response system, to direct its phase-space trajectory toward the trajectory of the drive system $\dot{\mathbf{v}} = \mathbf{F}(\mathbf{v})$. As soon as synchronization is achieved, the coupling vanishes. Note that not only scalar functions, but also vector couplings $\hat{\mathbf{C}}(\mathbf{u} - \mathbf{v})$ can be used (Brown et al. 1994). In this case, $\hat{\mathbf{C}}$ is a constant coupling matrix.

Stability of synchronization is defined by the conditional Lyapunov exponents calculated from variational equations of the response system. In the case of identical oscillators ($\mathbf{G} = \mathbf{F}$), we can write the linearized error equation as

$$\dot{\mathbf{e}} = [\mathbf{J}_F(\mathbf{v}) + \hat{\mathbf{C}}] \cdot \mathbf{e}, \tag{2.35}$$

where $\mathbf{e} = \mathbf{v} - \mathbf{u}$ is the synchronization error. The trajectory of the response system will be attracted to the trajectory of the drive if all Lyapunov exponents of Equation 2.35 are negative.

If the coupled oscillators are slightly different ($\mathbf{G} \approx \mathbf{F}$), the error dynamics can be described by the following linearized equation

$$\dot{\mathbf{e}} = \left(\mathbf{J}_F(\mathbf{v}) + \hat{\mathbf{C}}\right) \cdot \mathbf{e} - \mathbf{J}_{G-F}(\mathbf{v}) \cdot \mathbf{u}, \tag{2.36}$$

provided that **G** and **F** are smooth enough. When the coupling is strong enough, synchronization can be almost achieved, but it never appears under the form of complete synchronization. In other words, the magnitude of the synchronization error, $|\mathbf{e}|$, becomes very small, but it never vanishes.

The Fujisaka–Yamada scheme has received considerable attention because of its easy implementation in various systems, including electronic circuits and lasers, and was successfully used in chaotic communication (Hayes et al. 1993; Argyris et al. 2005; Pisarchik et al. 2012b; Pisarchik and Ruiz-Oliveras 2015).

2.4.4 Conditional Lyapunov Exponents for Discrete Systems

We now move to discuss conditional Lyapunov exponents for the case of two unidirectionally coupled one-dimensional discrete systems. Two variants of the Fujisaka–Yamada coupling scheme are commonly used.

Variant 1. The master (x) and slave (y) are coupled by their functions f as

$$\begin{aligned} x_{n+1} &= f_x(\mu_m, x_n), \\ y_{n+1} &= f_y(\mu_s, y_n) + \sigma(f_x(\mu_m, x_n) - f_y(\mu_s, y_n)), \end{aligned} \tag{2.37}$$

where μ_m and μ_s are parameters of the master and slave systems, σ is a coupling strength, and $n = 0, 1, 2, \ldots, N$ is the iteration number.

Variant 2. The systems are coupled by their variables as

$$x_{n+1} = f_x(\mu_m, x_n),$$
$$y_{n+1} = f_y(\mu_s, y_n) + \sigma(x_n - y_n). \tag{2.38}$$

These variants can be written in the following generic form:

$$x_{n+1} = f(x_n),$$
$$y_{n+1} = g(y_n) + \sigma(\varphi(x_n) - \psi(y_n)), \tag{2.39}$$

where g, φ, and ψ are functions of the slave map under the influence of the master map.

Thus, the coupled slave map can be presented as a system with an external drive $d(x_n) = \sigma\varphi(x_n)$ in the form

$$y_{n+1} = h(y_n) + d(x_n), \tag{2.40}$$

where $h(y_n) = g(y_n) - \sigma\psi(y_n)$ is a new function of the coupled slave map. The Jacobian matrix of Equation 2.39 is

$$\mathbf{J} = \begin{pmatrix} f' & 0 \\ \sigma\varphi' & g' - \sigma\psi' \end{pmatrix} = \begin{pmatrix} a_n & 0 \\ b_n & c_n \end{pmatrix}, \tag{2.41}$$

where the prime indicates derivative of the corresponding function, $a_n = f'(x_n)$, $b_n = \sigma\varphi'(x_n)$, and $c_n = h'(y_n) = g'(y_n) - \sigma\psi'(y_n)$. The product of the Jacobian matrices for the orbit $(x_1, y_1), \ldots, (x_N, y_N)$ is given by

$$\mathbf{J_N} \cdots \mathbf{J_2} \cdot \mathbf{J_1} = \begin{pmatrix} A_N & 0 \\ B_N & C_N \end{pmatrix}, \tag{2.42}$$

where $A_N = \prod_{n=1}^{N} a_n$, $B_N = \sum_{n=1}^{N} \left[b_n \left(\prod_{i=1}^{n-1} a_i \right) \left(\prod_{i=n+1}^{N} c_i \right) \right]$, and $C_N = \prod_{n=1}^{N} c_n$.

Due to the triangle structure of Equation 2.42, the conditional Lyapunov exponents λ_x and λ_y can be found as (Pisarchik et al. 2017)

$$\lambda_x = \lim_{N \to \infty} \frac{1}{N} \log |A_N| = \lim_{N \to \infty} \frac{1}{N} \sum_{n=1}^{N} \log |f'(x_n)|,$$

$$\lambda_y = \lim_{N \to \infty} \frac{1}{N} \log |C_N| = \lim_{N \to \infty} \frac{1}{N} \sum_{n=1}^{N} \log |g'(y_n) - \sigma\psi'(y_n)|. \tag{2.43}$$

In particular, for Variant 1, the functions of Equation 2.39 are $f(x) = f_x(\mu_m, x)$, $g(y) = f_y(\mu_s, y)$, $\varphi(x) = f_x(\mu_m, x)$, and $\psi(y) = f_y(\mu_s, y)$, and therefore the Lyapunov exponents are

$$\lambda_x = \lim_{N \to \infty} \frac{1}{N} \sum_{n=1}^{N} \log |f_x'(\mu_m, x_n)|,$$

$$\lambda_y = \lim_{N \to \infty} \frac{1}{N} \sum_{n=1}^{N} \log |f_y'(\mu_s, y_n) - \sigma f_y'(\mu_s, y_n)| \qquad (2.44)$$

$$= \log |1 - \sigma| + \lim_{N \to \infty} \frac{1}{N} \sum_{n=1}^{N} \log |f_y'(\mu_s, y_n)|.$$

For Variant 2, the functions in Equation 2.39 are $f(x) = f_x(\mu_m, x)$, $g(y) = f_y(\mu_s, y)$, $\varphi(x) = x$, and $\psi(y) = y$, and therefore the Lyapunov exponents are

$$\lambda_x = \lim_{N \to \infty} \frac{1}{N} \sum_{n=1}^{N} \log |f_x'(\mu_m, x_n)|,$$

$$\lambda_y = \lim_{N \to \infty} \frac{1}{N} \sum_{n=1}^{N} \log |f_y'(\mu_s, y_n) - \sigma|. \qquad (2.45)$$

To illustrate the above consideration, consider the simple example of the logistic map $f(\mu, x) = \mu x(1 - x)$ with parameter μ. In Figure 2.9 we plot the bifurcation diagrams of the slave logistic map versus the coupling σ for the two variants. The solid lines show the conditional Lyapunov exponents calculated by Equations 2.44 and 2.45. One can clearly see the existence of the periodic window for certain coupling strengths in both coupling configurations, while the uncoupled maps are chaotic. This effect, referred to as *coherence enhancement* in coupled oscillators, will be considered more explicitly in Section 2.7.

2.4.5 Example of Coupled Rössler-Like Oscillators

Next, we will discuss how synchronization arises in continuous time systems coupled in a master–slave configuration. Let us use the example of two chaotic Rössler-like oscillators evolving according to

$$\begin{aligned} \dot{x}_1 &= -\omega_1 y_1 - z_1, & \dot{x}_2 &= -\omega_2 y_2 - z_2, \\ \dot{y}_1 &= \omega_1 x_1 + a y_1, & \dot{y}_2 &= \omega_2 x_2 + a y_2 + \sigma(y_1 - y_2), \qquad (2.46) \\ \dot{z}_1 &= b + z_1(x_1 - c), & \dot{z}_2 &= b + z_2(x_2 - c). \end{aligned}$$

The drive system (x_1, y_1, z_1) is independent of the response system (x_2, y_2, z_2), while the latter depends on the variable y_1 of the drive. Here, ω_1 and ω_2 are the natural frequencies of the coupled oscillators.

We will first show how synchronization appears in the case of identical chaotic oscillators ($\omega_1 = \omega_2 = \omega = 1$), and then how it emerges in the presence of a small parameter mismatch $\Delta = \omega_2 - \omega_1$.

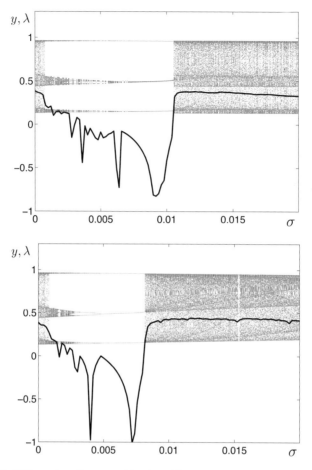

Figure 2.9 Bifurcation diagrams (dots) and Lyapunov exponents (solid lines) versus the coupling σ for a slave logistic map, unidirectionally coupled with a chaotic master logistic map in Variant 1 (upper panel) and Variant 2 (lower panel). The map parameters are $\mu_s = 3.853$ and $\mu_s = 3.860$.

The oscillators in Equation 2.46 are chaotic for $a = 0.16$, $b = 0.1$, $c = 8.5$, and $\omega = 1$, as seen from the time series and phase portrait in Figure 2.10. If the coupling strength is small enough, each oscillator, starting from different initial conditions, follows its own trajectory in the phase space, so that the synchronization error

$$|e| = \left| \sqrt{(x_2 - x_1)^2 + (y_2 - y_1)^2 + (z_2 - z_1)^2} \right| \qquad (2.47)$$

takes high values (Figure 2.10(b)). In the absence of synchronization, the phase-space portrait of the master and slave system variables occupies a large area as seen in Figure 2.10(c).

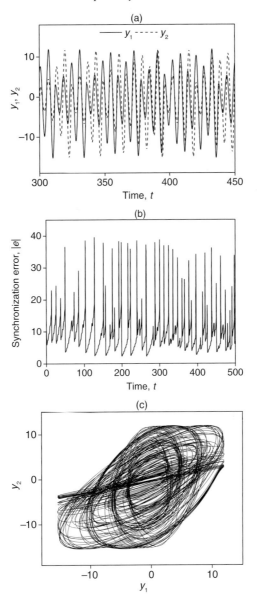

Figure 2.10 (a) Time series of variables y_1 and y_2, (b) synchronization error, and (c) phase-space portrait, of two uncoupled chaotic Rössler oscillators. See main text for the values of the parameters used. The synchronization error is very large.

Conversely, for sufficiently strong couplings ($\sigma > 0.5$), the master and slave oscillators become completely synchronized, meaning that their trajectories merge when $t \rightarrow \infty$ (Figure 2.11(a)), so that the synchronization error $|e| \rightarrow 0$ (Figure 2.11(b)), while the phase portrait becomes a straight line (Figure 2.11(c)).

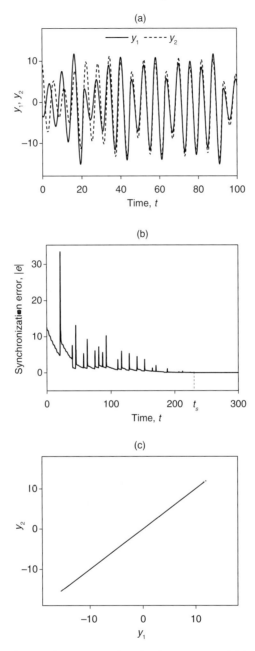

Figure 2.11 (a) Time series, (b) synchronization error, and (c) phase-space portrait after transients, of two chaotic Rössler oscillators coupled with $\sigma = 0.5$. The
synchronization error goes to zero when $t \to \infty$, meaning that the synchronization manifold is stable. The phase-space portrait of complete synchronization is a
straight line.

Looking at Figure 2.11(b) one immediately sees that the synchronization error is not always zero, but gradually decreases with time. The velocity of synchronization and the initial conditions determine the synchronization time t_s, which is another important measure of synchronization, and is defined as the time needed for the trajectory of a system to be constrained on the synchronization manifold.

The synchronization time depends on both the coupling strength and the initial conditions, with the latter dependence leading to uncertainties in the experimental estimation of t_s. In practical situations, one deals with *maximum synchronization time* and *mean synchronization time* as, respectively, the maximum and the most probable times needed by a system to synchronize when starting from a given (generic) initial condition. Since the condition for synchronization is $|e| \to 0$ when $t \to \infty$, the synchronization time is measured in practice as the time at which the error reaches a certain very small, but arbitrary, threshold magnitude. It has been shown that, in general, the synchronization time scales like $\sigma^{-1/2}$ (Sausedo-Solorio and Pisarchik 2011).

2.4.6 Stability of the Synchronization Manifold

As mentioned above, a necessary condition for synchronization is the stability of the synchronization manifold $\mathbf{x} \equiv \mathbf{y}$.

Since the error dynamics given by Equation 2.34 explicitly include the driving signal $\mathbf{s}(t)$, complete synchronization exists when the synchronization manifold is asymptotically stable for all possible trajectories $\mathbf{s}(t)$ of the driving system within the chaotic attractor.

This property can be proved via stability analysis of the linearized system for small error $\mathbf{e}$:

$$\dot{\mathbf{e}} = \mathbf{J_f}(\mathbf{s}(t)) \cdot \mathbf{e}, \tag{2.48}$$

where $\mathbf{J_f}(\mathbf{s}(t))$ is the Jacobian of the vector $\mathbf{f}$ evaluated onto the driving trajectory $\mathbf{s}(t)$. When the driving trajectory $\mathbf{s}(t)$ is a fixed point or a periodic orbit, the stability can be found from the eigenvalues of $\mathbf{J_f}$ or Floquet multipliers (Verhults 1990; Yu et al. 1990).

However, when the response system is driven by a chaotic signal, the stability can be evaluated by calculating the conditional Lyapunov exponents of the system Equation 2.48 defined as

$$\lambda(\mathbf{s}_0, \mathbf{u}_0) \equiv \lim_{t \to \infty} \frac{1}{t} \log \left(\frac{|\mathbf{e}(t)|}{|\mathbf{e}_0|} \right) = \lim_{t \to \infty} \frac{1}{t} \log |\mathbf{Z}(\mathbf{s}_0, t) \cdot \mathbf{u}_0| \tag{2.49}$$

for an initial condition of the driving signal $\mathbf{s}_0$ and initial orientation of the infinitesimal displacement $\mathbf{u}_0 = \mathbf{e}(0)/|\mathbf{e}(0)|$. Here, $\mathbf{Z}(\mathbf{s}_0, t)$ is the matrix solution of the linearized equation

$$\dot{\mathbf{Z}} = \mathbf{J_f}(\mathbf{s}(t)) \cdot \mathbf{Z} \tag{2.50}$$

subject to the initial condition $\mathbf{Z}(\mathbf{0}) = \mathbf{I}$. The synchronization error $\mathbf{e}$ evolves according to $\mathbf{e}(t) = \mathbf{Z}(\mathbf{s}_0, t) \cdot \mathbf{e}_0$, and then the matrix $\mathbf{Z}$ determines whether this error shrinks or grows in a particular direction.

Usually, the conditional Lyapunov exponents are calculated numerically from a temporal average, and therefore they characterize the global stability over the whole chaotic attractor. It should be noted that the negativity of the conditional Lyapunov exponents is only a necessary condition for the stability of the synchronization manifold. In some cases, additional stability conditions must be satisfied to warrant synchronization in a sufficient way (Willems 1970).

Another way to study stability of a synchronization manifold is the method based on the Lyapunov function (Kocarev and Parlitz 1995; Cuomo and Oppenheim 1993), which instead yields both necessary and sufficient conditions. Using Equation 2.34, the *Lyapunov function* $\Lambda(\mathbf{e})$ is defined as a continuously differentiable real valued function with the following properties:

$$\text{(i)} \quad \Lambda(\mathbf{e}) > 0 \text{ for all } \mathbf{e} \neq 0 \text{ and } \Lambda(\mathbf{e}) = 0 \text{ for } \mathbf{e} = 0, \tag{2.51}$$

$$\text{(ii)} \quad \frac{d\Lambda}{dt} < 0 \text{ for all } \mathbf{e} \neq 0. \tag{2.52}$$

If the Lyapunov function exists, then the complete synchronization manifold is globally stable for the coupled system.

One more criterion for the stability of synchronized states was introduced by Brown and Rulkov (1997a; 1997b). They divided the synchronization error dynamics into two parts, a time-independent part $\mathbf{A}$ and a time-dependent part $\mathbf{B}(\mathbf{x}, t)$, so that

$$\dot{\mathbf{e}} = (\mathbf{A} + \mathbf{B}(\mathbf{x}, t)) \cdot \mathbf{e}. \tag{2.53}$$

If we assume that $\mathbf{A}$ can be diagonalized, then, transforming $\mathbf{B}$ into the coordinate system defined by the eigenvectors of $\mathbf{A}$, a sufficient condition for the stability of the synchronization manifold is

$$-\Re(\lambda_m) > \langle \|\mathbf{P}^{-1}\mathbf{B}\mathbf{P}\| \rangle. \tag{2.54}$$

Here, $\Re(\lambda_m)$ is the real part of the largest eigenvalue of $\mathbf{A}$, $\mathbf{P}$ is the matrix of eigenvectors of $\mathbf{A}$, the angled brackets denote a time average along the driving trajectory, and $\| \cdot \|$ indicates the Frobenius norm, defined as

$$\|\mathbf{M}\| = \sqrt{\text{Tr}(\mathbf{M}^\dagger \mathbf{M})},$$

where Tr is the trace operator, and $\dagger$ denotes Hermitian conjugate.

Since this condition is based on a matrix norm, it is only sufficient: a driving configuration may well not satisfy this condition, and yet it may produce a stable synchronized motion.

Only Lyapunov functions give a necessary and sufficient condition for the stability of the synchronization manifold, while the negativity of the conditional Lyapunov exponents and the criteria of Equation 2.54 provide respectively a necessary condition only and a sufficient condition only.

In particular, the Lyapunov function criterion gives a local condition for stability involving temporal averages over chaotic trajectories of the driving signal, and therefore establishes conditions for global stability. As a consequence, it is not affected by local desynchronization events that could occur within the synchronization manifold.

2.5 Phase Synchronization

Phase synchronization frequently occurs in coupled systems and plays a crucial role in many weakly interacting ones, including lasers (Pisarchik 2008), electronic circuits (Parlitz et al. 1996b; Carroll 2001; Roy et al. 2003), cardiorespiratory rhythm (Schäfer et al. 1998), neurons (Makarenko and Llinás 1998; Tass et al. 1998), human behavior (Bhattacharya and Petsche 2001), magnetic and electric brain dynamics (Tass et al. 1998), and ecological systems (Blasius et al. 1999; Blasius and Stone 2000).

Phase synchronization is particularly relevant for biomedical applications, as it provides a measure of coordinated electromagnetic activity of different brain areas, and describes long-range synchronization patterns in separated brain regions that are involved with cognitive mechanisms, memory, emotions, and motor panning.

This phenomenon takes place when the phase difference between chaotic oscillations is asymptotically bounded within 2π. This is the weakest manifestation of synchronization in chaotic systems, and it should not be confused with antiphase synchronization, which is instead a much stronger type of synchronization defined as the state when the variables of two interacting systems have the same amplitude but differ in sign. Historically, antiphase synchronization was the first type of synchronization observed experimentally by Huygens in coupled pendula, with experiments that have also been reproduced using modern setups and technologies (Bennet et al. 2002).

2.5.1 Defining Phases in Chaotic Systems

While instantaneous phases can always be well defined in periodic systems, their definition in chaotic systems is somewhat problematic. The most popular approaches to associate an *instantaneous* phase to a chaotic trajectory are the introduction of geometric phases, the use of the Poincaré section and Hilbert transform, and the definition of phase coordinates. It is therefore convenient to briefly review these methods, which provide a necessary first step toward describing and measuring phase synchronization.

Geometric Phase

Chaotic systems for which a given center of rotation exists such that the trajectory rotates around it are called *phase-coherent*. Phase-incoherent systems are those, instead, for which a center of rotation cannot be defined, in the sense that there is a set of points, around each one of which the chaotic trajectory performs at least one rotation, or (equivalently) for any point in phase space the trajectory performs at least one rotation that is not centered around it.

In phase-coherent systems, one can define an instantaneous phase as the angle

$$\varphi = \arctan\left(\frac{y(t) - y_0}{x(t) - x_0}\right),$$

that the chaotic trajectory, projected on a given $(x - y)$ plane of the phase space, forms during its rotation around the point (x_0, y_0).

For two uncoupled chaotic, phase coherent, systems, the magnitude of the difference $\vartheta(t) = \varphi_1(t) - \varphi_2(t)$ between the instantaneous phases increases linearly with time. The so-called $n : m$ phase synchronization (with n and m being arbitrary integer numbers) manifests itself when the *phase-locking condition*

$$|\vartheta(t)| = |n\varphi_1(t) - m\varphi_2(t)| < c \tag{2.55}$$

is satisfied asymptotically, for a constant c.

In coupled nonidentical oscillators, phase locking is accompanied by frequency entrainment, i.e., the frequencies $n\Omega_1$ and $m\Omega_2$ coincide. The *frequency entrainment condition* can be written as follows

$$|\Delta(t)| = |n\dot\varphi_1(t)| - |m\dot\varphi_2(t)| = 0, \tag{2.56}$$

where Δ stands for the difference between the dominant frequencies (multiplied by the respective integers n and m).

Hilbert Transform

The phase variables $\varphi_1(t)$ and $\varphi_2(t)$ of two chaotic systems can also be obtained by means of an analytic signal approach (Gabor 1946; Prigogine and Stengers 1965). There, the instantaneous phase of an arbitrary signal $s(t)$ is defined as

$$\varphi(t) = \arctan\frac{\bar{s}(t)}{s(t)}, \tag{2.57}$$

where

$$\bar{s}(t) = \frac{1}{\pi} \, \text{p.v.} \int_{-\infty}^{\infty} \frac{s(\tau)}{t - \tau} d\tau \tag{2.58}$$

is the Hilbert transform of the observed scalar time series $s(t)$ (p.v. denoting the Cauchy principal value).

In other words, $\bar{s}(t)$ is the convolution of $s(t)$ with $1/(\pi t)$. Since the Fourier transform of a convolution of functions is the product of the Fourier transforms of the two functions, one has

$$\mathcal{F}(\bar{s}(t)) = \mathcal{F}(s(t)) \, \mathcal{F}\left(\frac{1}{\pi t}\right) = -\frac{i}{\sqrt{2\pi}} \text{sign}(\omega) \, \mathcal{F}(s(t)), \tag{2.59}$$

where ω is the frequency variable of the Fourier domain, and the Fourier transform of $1/(\pi t)$ is itself computed using the Cauchy principal value. From this, it follows that

$$\bar{s}(t) = \mathcal{F}^{-1}\left(-\frac{i}{\sqrt{2\pi}} \text{sign}(\omega) \, \mathcal{F}(s(t))\right). \tag{2.60}$$

This last equation can be rewritten as

$$\bar{s}(t) = \begin{cases} \mathcal{F}^{-1}\left(-\frac{i}{\sqrt{2\pi}} \mathcal{F}(s(t))\right) & \text{for } \omega > 0 \\ \mathcal{F}^{-1}\left(\frac{i}{\sqrt{2\pi}} \mathcal{F}(s(t))\right) & \text{for } \omega < 0. \end{cases} \tag{2.61}$$

From here, it becomes clear that the Hilbert transform performs a phase shift of the original signal by $\pi/2$ for negative frequencies and by $-\pi/2$ for positive ones.

It is important to note that the instantaneous phase defined in Equation 2.57 is restricted to the interval $[0, 2\pi]$ and has to be unfolded or continued, i.e., shifted by 2π whenever a 2π-phase slip occurs, prior to taking its derivative (cf. Equation 2.56). Due to these phase jumps, the dominant frequency is strongly affected by noise which makes it an inadequate measure for synchronization in noisy time series such as, for instance, electroencephalograms (EEG) and magnetoencephalograms (MEG).

To overcome this problem, a more statistical approach can be used: one can analyze the distribution of relative phase angles on the unit circle, i.e., the interval $[0, 2\pi]$. If the phases are locked during most of the time, a prominent peak will result in the phase histogram, and the effect of 2π-phase jumps will no longer be dominant.

When using the analytic signal approach, one way of confining the phase difference to the interval $[0, 2\pi]$ is applying a trigonometric addition theorem. For 1:1 phase locking, this yields

$$\vartheta^{1:1}(t) = \arctan \frac{\bar{s}_1(t)s_2(t) - s_1(t)\bar{s}_2(t)}{s_1(t)s_2(t) + \bar{s}_1(t)\bar{s}_2(t)}. \tag{2.62}$$

Poincaré Section

The phase associated with a chaotic time series can also be measured based on an appropriate Poincaré section, that the chaotic orbit intersects in one direction at each rotation.

Successive crossings of the Poincaré section can indeed be associated with a phase increase of 2π, and all values of the instantaneous phase between one crossing and the next can be computed by linear interpolation as

$$\varphi(t) = 2\pi k + 2\pi \frac{t - t_k}{t_{k+1} - t_k} \qquad (t_k < t < t_{k+1}), \qquad (2.63)$$

where t_k is the time at which the kth crossing occurs.

This phase definition is very useful for practical applications when only one system variable is available for measurement, because it allows one to obtain an estimation simply from the scalar chaotic time series, computing the t_k from the times of successive local maxima or minima, without having to reconstruct a higher-dimensional phase space to explicitly find a Poincaré section.

This technique is particularly appealing for the analysis of phase synchronization in non-smooth systems, such as, for example, a piecewise linear Rössler oscillator (Pisarchik et al. 2006), as will be discussed later on.

Phase Coordinate

Another practical definition of the phase of chaotic oscillators can be made by studying the influence of a small periodic perturbation on a chaotic system (Josić and Mar 2001).

To that end, we start by considering the simple case of a periodic dynamical system

$$\dot{\mathbf{f}} = \mathbf{f}(\mathbf{x}) \qquad (2.64)$$

with a flow φ_t. Assuming that this system has a stable limit cycle ρ of period T, one will find coordinates $(\varphi, \mathbf{R})$ in a neighborhood N of ρ such that the phase φ is the angular distance along ρ, $\mathbf{R}$ is the radial distance from ρ, and $d\varphi/dt = 1$.

Every level set of φ, called an *isochron*, is a codimension-one manifold that foliates N and intersects ρ in a point q_φ referred to as the *basepoint* of the isochron. There exist $C > 0$ and $k > 0$ such that, for any point p on an isochron with basepoint q_φ, the condition

$$|\varphi_t(p) - \varphi_t(q_\varphi)| \leqslant C e^{kt}$$

is satisfied. This means that the asymptotic behavior of all points on an isochron is the same as that of its basepoint.

Now, let the system Equation 2.64 be subjected by a small perturbation $\varepsilon\mathbf{p}(t)$ of period T_d so that the perturbed system

$$\dot{\mathbf{f}} = \mathbf{f}(\mathbf{x}) + \varepsilon\mathbf{p}(t)$$

possesses an attracting limit cycle ρ_ε which is $\mathcal{O}(\varepsilon)$ close to ρ. Direct calculation yields

$$\dot{\varphi} = 1 + \varepsilon\Omega(\varphi, t) + \mathcal{O}(\varepsilon^2),$$

where the value

$$\Omega(\varphi, t) \equiv \nabla_{\mathbf{x}}\varphi|_{\rho(\varphi)} \cdot \mathbf{p}(t)$$

measures the influence of the perturbation on the phase. Here, $\nabla_{\mathbf{x}}\varphi|_{\rho(\varphi)}$ is the gradient of $\varphi(\mathbf{x})$ evaluated at the point $\rho(\varphi)$ of the perturbed orbit, and is interpreted as the phase-dependent sensitivity of φ.

Defining the phase difference between the perturbation $\mathbf{p}(t)$ and the phase φ as

$$\Psi = \varphi - \frac{T}{T_d}t,$$

its dynamics can be described by

$$\dot{\Psi} = \varepsilon\left[\Delta + \Omega\left(\frac{T}{T_d}t + \Psi, t\right)\right], \tag{2.65}$$

where

$$\Delta \equiv \frac{1}{\varepsilon}\left(1 - \frac{T}{T_d}\right).$$

Averaging the phase difference over one period of the drive using the function

$$\Gamma(\Psi) = \frac{1}{T_d}\int_0^{T_d}\Omega\left(\frac{T}{T_d}t + \Psi, t\right)dt,$$

Equation 2.65 becomes

$$\dot{\Psi} = \varepsilon\left[\Delta + \Gamma(\Psi)\right]. \tag{2.66}$$

If Equation 2.66 has a stable fixed point Ψ_0, we say that the system Equation 2.64 is phase locked with the drive with a phase difference Ψ_0, i.e., the phase φ approaches the solution $\varphi(t) = \varphi(T + T_d)$.

Josić and Mar (2001) have extended the above consideration to the case of chaotic systems. They assumed that there exist coordinates of radial distance $\mathbf{R}$ and T-periodic phase φ in the neighborhood of the chaotic attractor, such that

$$\dot{\mathbf{R}} = \mathbf{f}(\mathbf{R}, \varphi),$$
$$\dot{\varphi} = 1 + \delta(\mathbf{R}, \varphi). \tag{2.67}$$

With this coordinate transformation, the phase dynamics are similar to those of a periodic orbit, with the exception of the term $\delta(\mathbf{R}, \varphi)$ indicating the phase sensitivity to the amplitude $\mathbf{R}$, which is required to be small, i.e., $\delta(\mathbf{R}, \varphi)$ is $\mathcal{O}(\varepsilon)$ where $\varepsilon \ll 1$. Furthermore, two points (R_1, φ) and (R_2, φ), starting from the same initial phase, remain close in phase for times at least $\mathcal{O}(\varepsilon)$ before they are separated due to the effect of the term δ. This means that, up to order ε, the level sets of φ form isochrons, and that the system is *phase-coherent*.

Although such a change of coordinates does not always exist, it is often possible to define a phase Φ and a natural period T such that $|\Phi(T) - \Phi(0)| < \varepsilon \ll 1$ by using the Hilbert transform. The phase fluctuation in time due to the term δ is known as *phase diffusion*.

It is also important to point out that the phase of a chaotic flow is closely related to its associated zero Lyapunov exponent (Rosenblum et al. 1996). The zero Lyapunov exponent corresponds, in fact, to the translation $\dot{\mathbf{x}}(t)$ along the chaotic trajectory. In a system where the chaotic flow has a proper rotation around a certain reference point, $\dot{\mathbf{x}}(t)$ can be uniquely mapped to a shift of the oscillator phases. For this reason, phase synchronization of chaotic oscillators can be revealed by a transition in the zero Lyapunov exponent.

2.5.2 Measures of Phase Synchronization

Having discussed the most common methods of defining instantaneous phases for chaotic systems, we can now move on to describing how one can actually measure the level of phase synchronization between two coupled systems.

We focus here on four indices used for this purpose, namely, the phase-locking value, the Shannon entropy index, the first Fourier mode index, and the conditional probability index.

Phase-Locking Value

For the analysis of biological time series, such as EEG and MEG, phase synchronization is often estimated with the so-called phase-locking value.

To define this quantity, one starts by mapping the relative phase angles of the oscillators onto the unit circle in the complex plane. Then, for a data set with N samples taken at regular time intervals τ, the phase-locking value R is defined as

$$R = \left| \frac{1}{N} \sum_{j=0}^{N-1} e^{i\vartheta(j\tau)} \right|. \tag{2.68}$$

Note that this is just the length of the circular mean of the sampled data (Mardia 1972).

The use of Euler's formula turns Equation 2.68 into

$$R = \frac{1}{N} \sqrt{\left[\sum_{j=0}^{N-1} \sin\left(\vartheta\left(j\tau\right)\right) \right]^2 + \left[\sum_{j=0}^{N-1} \cos\left(\vartheta\left(j\tau\right)\right) \right]^2}. \qquad (2.69)$$

From this notation, it becomes evident that R, being restricted to the interval $[0, 1]$, reaches the value 1 if and only if the condition of strict phase locking is fulfilled, whereas a uniform distribution of phases, which would be expected, on average, for unsynchronized time series, results in $R = 0$.

Shannon Entropy Index

In the analysis of physiological data, the Shannon entropy is often used to define an index of phase synchronization, useful to determine the existence of an $n : m$ phase-locking ratio.

This can be done by comparing the distribution of the cyclic relative phase $\varphi_{n,m} = (n\varphi_1 - m\varphi_2) \mod 2\pi$ with the uniform distribution. To do so, one first builds a histogram of $\varphi_{n,m}$ over N bins. Calling p_k the fraction of values of $\varphi_{n,m}$ that fall within the kth bin, the entropy of the distribution of $\varphi_{n,m}$ is

$$S = -\sum_{k=1}^{N} p_k \log p_k. \qquad (2.70)$$

The synchronization index is then defined as

$$\rho_{n,m} = \frac{\log\left(N - S\right)}{\log N} = 1 - \frac{S}{\log N}. \qquad (2.71)$$

For L data points, the optimal number of bins is $N_{opt} = e^{0.626 + 0.4\log(L-1)}$ (Otnes and Enochson 1972).

The synchronization index of Equation 2.71 is bounded between 0 and 1. When $\rho_{n,m} = 0$, the distribution is uniform, indicating the absence of synchronization, whereas $\rho_{n,m} = 1$ corresponds to a delta Dirac distribution, indicating complete synchronization. Note that the latter is only observed in case of ideal phase locking of noiseless oscillators.

First Fourier Mode Index

A further index, typically used in the same cases as the Shannon entropy index, is based on the intensity of the first Fourier mode of the distribution of $\varphi_{n,m}$, and it is defined as

$$\gamma_{n,m} = \sqrt{\left\langle \cos\left(\varphi_{n,m}\right)\right\rangle^2 + \left\langle \sin\left(\varphi_{n,m}\right)\right\rangle^2}, \qquad (2.72)$$

where the angled brackets denote time average. Also this index can vary between 0 (no synchronization) and 1 (perfect $n : m$ phase synchronization), and its advantage is that one does not need to calculate the distribution of the cyclic relative phase itself, but only its first Fourier mode.

The first Fourier mode index is very convenient for estimating phase synchronization from physiological signals. For example, by the use of this index, Gong et al. (2007) found collective phase synchronization in brain spontaneous activity in the alpha range (8–13 Hz) of the human EEG. However, since physiological data are very noisy and phase slips frequently occur, the relative phase difference $\psi_k(t, a) = |\varphi_k(t, a) - \varphi_l(t, a)|$ between the signals from the kth and lth channels fluctuates around a certain constant value. Therefore, phase synchronization can only be treated in a statistical sense. Using a sliding window of length n, one can calculate the first Fourier mode index for phase synchronization between the kth and lth channels as follows:

$$\gamma_{k,l} = \sqrt{\left[\frac{1}{n+1}\sum_{j=-n/2}^{n/2}\sin\left(\psi_{k,l}(t+j\tau, a)\right)\right]^2 + \left[\frac{1}{n+1}\sum_{j=-n/2}^{n/2}\cos\left(\psi_{k,l}(t+j\tau, a)\right)\right]^2},$$

where τ is the sampling time step and $j \in [-n/2, n/2]$ is an integer.

In Figure 2.12 we illustrate how synchronization appears in two unidirectionally coupled identical chaotic Rössler oscillators (Equation 2.46), when the coupling strength is increased. One can see from Figure 2.12 that synchronization index γ (calculated with Equation 2.72) gradually increases, approaching 1, while the synchronization error $|e|$ (calculated with Equation 2.47) gradually decreases, approaching 0, as seen from Figure 2.11(b).

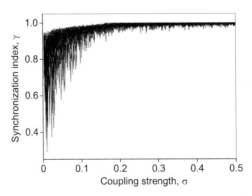

Figure 2.12 Synchronization index as a function of the coupling strength in the coupled chaotic Rössler oscillators of Equation 2.46 for $a = 0.16$, $b = 0.1$, $c = 8.5$, and $\omega = 1$.

The variation of both γ and $|e|$ is determined by the drift of the phase difference due to imperfect phase synchronization. The amplitude of the phase drift characterizes the quality of phase synchronization; the smaller the drift, the better the synchronization.

Conditional Probability Index

A final index, often used in the same circumstances as the previous two, is based on conditional probability. It is especially useful to measure synchronization of strongly nonlinear oscillators whose distribution of $\varphi_{n,m}(t)$ is nonuniform even in the absence of noise. This index is defined as

$$\eta = \varphi_2 \bmod 2\pi n|_{\varphi_1 \bmod 2\pi m = \vartheta}. \tag{2.73}$$

This means that the phase of one of the coupled oscillators (φ_2) is measured at the instants of time when the phase of another oscillator (φ_1) attains a certain fixed value ϑ (phase stroboscope).

To account for the $n : m$ locking, the phases of the oscillators 1 and 2 are wrapped into intervals $[0, 2\pi m]$ and $[0, 2\pi n]$, respectively. Repeating this procedure for all phases $\varphi \in [0, 2\pi]$ and averaging, one gets a statistically significant synchronization index.

In practice, when dealing with time series, one introduces a binning for the phase oscillators. Then, one defines $\eta_{i,j}$ as the conditional probability that $\varphi_2 \bmod 2\pi n$ is within the ith bin when $\varphi_1 \bmod 2\pi m$ is within the jth bin. If the oscillators are not synchronized, the $\eta_{i,j}$ are uniformly distributed; otherwise their distribution is unimodal for any choice of j.

2.5.3 Example: Ring of Rössler Oscillators

To describe how phase synchronization arises in nonidentical chaotic systems, we start by considering a ring of three unidirectionally coupled Rössler oscillators (Figure 2.13), with a small mismatch between their natural frequencies $\omega_1 < \omega_2 < \omega_3$. The ring is, indeed, one of the simplest motifs, which may repeat in a specific network or even among various networks, and can be responsible for particular

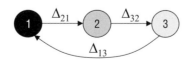

Figure 2.13 Ring of three unidirectionally coupled oscillators. The links among the oscillators are here labeled with the associated frequency mismatch $\Delta_{ji} = \omega_j - \omega_i$.

functions. As a consequence, understanding its synchronization properties is very important for the more general knowledge of the dynamical behavior of complex networks.

The oscillators are coupled through x variable, so that their dynamics are described by the following system of equations

$$
\begin{aligned}
\dot{x}_j &= -\omega_j y_j - z_j + \sigma(x_i - x_j), \\
\dot{y}_j &= \omega_j x_j + a y_j, \\
\dot{z}_j &= b + z_j(x_j - c),
\end{aligned} \tag{2.74}
$$

where $i, j = 1, 2, 3$ ($i \neq j$) is the number of the master i and slave j oscillator, $\omega_1 = 0.95$, $\omega_2 = 0.97$ and $\omega_3 = 0.99$ are the oscillators' natural frequencies, and σ is the coupling strength.

When the oscillators are uncoupled ($\sigma = 0$), all of them give rise to chaotic dynamics for the choice of parameters $a = 0.165$, $b = 0.2$, and $c = 10$. Starting from different initial conditions, they oscillate asynchronously, as can be seen from the time series presented in Figure 2.14(a), and the corresponding power spectra of the x variables shown in Figure 2.14(b). The spectra, furthermore, exhibit maxima at dominant frequencies $\Omega_1^0 \approx 0.975$, $\Omega_2^0 \approx 0.998$, and $\Omega_3^0 \approx 1.02$. All these frequencies are a bit different from the natural frequencies ω_i of the corresponding oscillators, because of the nonlinear character of the evolution equations, which actually couples the amplitude and the frequency of oscillations (Nayfeh and Mook 1979). In the case considered, they are all tuned towards higher frequencies.

In the following, we describe the scenario that emerges as the coupling strength σ is increased. As we already mentioned, one possibility for monitoring phase synchronization between a pair of phase coherent oscillators j and i that rotate around a center (x_0, y_0) by means of the difference between their instantaneous phases $\vartheta_{ji} = \varphi_j - \varphi_i$, where $\varphi = \arctan\left[(y - y_0) / (x - x_0)\right]$.

Since the oscillators have different natural frequencies, ϑ_{ji} changes monotonically with time when they are uncoupled, increasing or decreasing depending on the sign of the frequency mismatch. Already for a very small value of the coupling strength ($\sigma \geq 5 \times 10^{-3}$), irregular windows of phase-synchronized motion (laminar phases) appear in the time series. These windows are interrupted by sudden phase slips (turbulent phases) during which the phase difference $|\vartheta_{ji}(t)|$ jumps up by 2π. Such a regime is referred to as *intermittent phase synchronization*. During the laminar phases, a master oscillator locks the phase of the corresponding slave oscillator, so that their dominant frequencies match. As a consequence, $\vartheta_{ji}(t)$ neither increases nor decreases monotonically in time, but fluctuates around a certain average value.

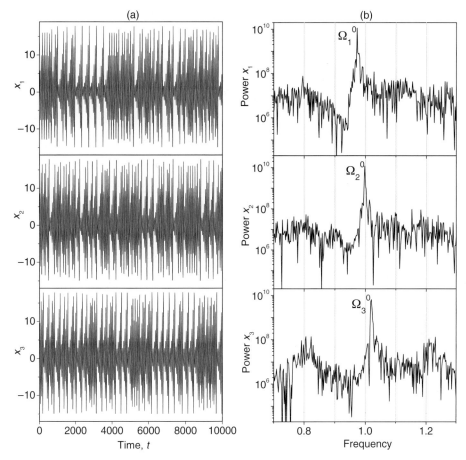

Figure 2.14 (a) Time series and (b) power spectra of x variable of three uncoupled Rössler oscillators demonstrating asynchronous chaotic behavior with dominant frequencies $\Omega_1^0 = 0.975$, $\Omega_2^0 = 0.998$, and $\Omega_3^0 = 1.02$.

This scenario is illustrated in Figure 2.15, where the time evolutions of ϑ_{21} are reported for three different coupling strengths. The segments in time where the behavior of ϑ_{21} is almost aligned with the horizontal axis correspond to the windows of phase synchronization, where the dominant frequency of the slave oscillator is locked by the corresponding master oscillator.

A gradual increase in σ progressively enlarges these windows, leading eventually (i.e., for sufficiently high values of σ) to permanent phase synchronization. Thus, the slope α of the fitting straight line (or time-averaged phase) is a good quantitative measure of intermittent phase synchronization; the smaller is α, the larger is the duration of the phase synchronization windows. A vanishing slope ($\alpha = 0$) indicates permanent phase synchronization.

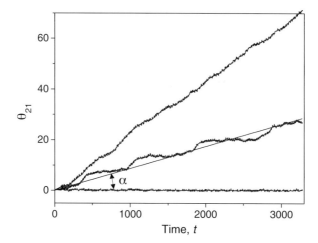

Figure 2.15 Time behavior of phase difference ϑ_{21} between oscillators 2 and 1 for $\sigma = 6.6 \times 10^{-3}$ (upper line), 2.6×10^{-2} (middle line), and 4.6×10^{-2} (lower line). The epochs in which the time evolution is almost aligned with the horizontal axis are the windows of phase synchronization, which alternate with asynchronous motion. The straight line with slope α is a linear fit of the long-term average phase dependence.

Note that, although the phase difference fluctuations are generated by fully deterministic equations, they look noisy. Indeed, for weak coupling strengths, they obey the same scaling relations as a Brownian motion, whereas for stronger coupling they behave as pink or $1/f$ noise. In coupled chaotic systems, chaos acts in a similar way as phase noise in a signal generator, broadening its power spectrum.

Due to the phase locking, the dominant frequency of the slave oscillator is the same as that of the corresponding master oscillator in the phase synchronization windows, whereas out of these windows the oscillators have distinct dominant frequencies.

Intermittent phase synchronization can be also characterized by the time-averaged dominant frequency $\langle \Omega_i \rangle$; the closer the average dominant frequencies of coupled oscillators are to each other, the better the synchronization. Figure 2.16 shows how both the time-averaged dominant frequency $\langle \Omega_i \rangle$ of all oscillators and the slopes α_{ji} of the time-dependent phase differences depend on the coupling strength.

As σ increases, $\langle \Omega_2 \rangle$ and $\langle \Omega_3 \rangle$ slowly decrease (see Figure 2.16(a)) because the duration of intermittent frequency-locked windows increases. Meanwhile, almost no changes occur in $\langle \Omega_1 \rangle$ because the distance (or frequency mismatch) between the oscillators 3 and 1 is too large.

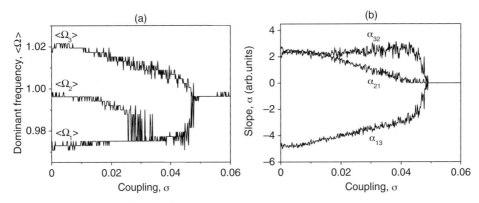

Figure 2.16 (a) Time-averaged dominant frequencies $\langle\Omega_i\rangle$ and (b) slopes α_{ji} as a function of the coupling strength σ.

Only when σ approaches a threshold value ($\sigma_{th} \approx 0.04$), oscillator 1 starts to intermittently change its dominant frequency, and its time-averaged dominant frequency $\langle\Omega_1\rangle$ grows up very fast as σ is further increased. Eventually, for strong enough coupling ($\sigma > 0.05$) the dominant frequencies of all three oscillators are completely locked, which results in phase synchronization.

The slopes α_{ji} are shown in Figure 2.16(b): initially oscillators 1 and 2 synchronize with each other, and then they both synchronize with the remaining oscillator 3.

One can see from Figure 2.16 that for every pair of oscillators i, j there exist two characteristic values of the coupling strength σ_1 (i, j) and σ_2 (i, j) that separate different slopes of the curves. The lower threshold, σ_1 (i, j), indicates the onset of intermittent phase synchronization, while for couplings greater than the higher threshold σ_2 (i, j), phase synchronization takes place between i and j.

When discussing intermittent phase synchronization, a relevant question is what type of intermittency such a phenomenon belongs to. After its discovery in the vicinity of the phase synchronization boundary, it was classified as *eyelet intermittency* (Pikovsky et al. 1997; Lee et al. 1998; Rosa et al. 1998). This type of intermittency is characterized by the exponential dependence of the mean length T of the laminar phases on the criticality parameter σ_2 given by

$$\log(1/T) = c_0 - c_1(\sigma_2 - \sigma)^{-1/2},$$

where c_0 and c_1 are constants (Grebogi et al. 1983). Eyelet intermittency appears due to boundary crisis of the synchronous attractor caused by an unstable–unstable bifurcation when a saddle periodic orbit and a repeller periodic orbit join and die (Pikovsky et al. 1997; Rosa et al. 1998).

Later, when eyelet intermittency was analyzed from the viewpoint of noise-induced intermittency, it was surprisingly found that it behaves as noise-induced type-I intermittency (Kye and Kim 2000; Hramov et al. 2011). The type-I intermittency is known to be observed below a saddle-node bifurcation and characterized by the scaling law

$$T \sim (\varepsilon_c - \varepsilon)^{-1/2},$$

where ε is a control parameter and ε_c is its bifurcation value (Pomeau and Manneville 1980).

In the presence of noise, the position of the saddle-node bifurcation depends on the noise intensity, so that the system near the onset of the type-I intermittency changes its behavior below and above the saddle-node bifurcation point ε_c (Eckmann 1981). Above this bifurcation ($\varepsilon > \varepsilon_c$), the mean length T of the laminar phase obeys the law (Hramov et al. 2007):

$$T = \frac{1}{k\sqrt{\varepsilon - \varepsilon_c}} e^{\frac{4(\varepsilon-\varepsilon_c)^{3/2}}{3D}}, \qquad (2.75)$$

where k is a constant and D is the intensity of delta-correlated white noise. If the criticality parameter ε is large enough, the approximate equation

$$\log T \sim D^{-1}(\varepsilon - \varepsilon_c)^{3/2}$$

can be used (see, for example, Kye and Kim 2000).

The coincidence of two different types of intermittency is understandable if one assumes that the fluctuations of the phase difference behave as a random walk, similar to a Brownian motion (Pisarchik et al. 2011).

In summary, the system evolution towards phase synchronization, as the coupling strength is increased, obeys the following scenario:

(i) Intermittent phase-locking with a drift of the phase difference $\vartheta(t) \in [-\pi, \pi]$. In the case of more than two coupled oscillators, the phase is intermittently locked.
(ii) Imperfect phase-locking with decreasing range of the phase drift.
(iii) Perfect phase-locking with almost zero phase drift.

In the next subsection we will describe what happens for a further increase in the coupling strength.

2.6 Lag and Anticipating Synchronization

Lag and anticipating synchronization are two specific forms of synchronization occurring for values of the coupling strength generally higher than those giving rise to phase synchronization. These types of synchronization are observed in two cases: (i) when the coupling between the subsystems includes some time delay (Voss 2000, 2001), and (ii) in the presence of a mismatch between the natural frequencies of the coupled oscillators (Corron et al. 2005; Pyragiené and Pyragas 2013).

Lag synchronization arises as an intermediate step in the route from phase synchronization to complete synchronization, i.e., as an initial step where amplitudes and phases of the chaotic systems start to display some form of functional correlation.

We first consider case in which there is a propagation delay in sending the coupling signal from the drive to the response. If the chaotic drive oscillator

$$\dot{\mathbf{x}} = \mathbf{F}(\mathbf{x})$$

is coupled with the chaotic response oscillator by means of a propagation delay, the response system in lag synchronization can be written as

$$\dot{\mathbf{y}}_l = \mathbf{F}(\mathbf{y}_l) + \mathbf{G}\left[\mathbf{x}(t - \tau), \mathbf{y}_l(t)\right], \tag{2.76}$$

where $\mathbf{y}_l(t)$ is the response state and τ is the propagation delay.

A solution to Equation 2.76 is the time-shifted synchronous state

$$\mathbf{y}_l(t) = \mathbf{x}(t - \tau),$$

where the response system exactly follows the drive with a time lag $\tau > 0$. A somehow more surprising situation is that of anticipating synchronization, which can be achieved in a response system of the type

$$\dot{\mathbf{y}}_a = \mathbf{F}(\mathbf{y}_a) + \mathbf{G}\left[\mathbf{x}(t + \tau), \mathbf{y}_a(t)\right]. \tag{2.77}$$

For the system in Equation 2.77, the time-shifted synchronous state

$$\mathbf{y}_a(t) = \mathbf{x}(t + \tau)$$

is also a solution, but now the response system is shifted forward in time with respect to the drive, i.e., it synchronizes with a future state of the drive or, in other words, it *anticipates* the drive.

Alternatively, lag and anticipating responses can be obtained without an explicit delay in the coupling. To demonstrate it, let us consider the anticipating response Equation 2.77, for which the time-shifted synchronous state $\mathbf{y}_a(t)$ is the exact solution. Performing a time translation $t \to t - \tau$, for small τ, the coupling function can be linearly approximated as

$$\mathbf{G}\left[\mathbf{x}(t), \mathbf{y}_a(t - \tau)\right] \approx \mathbf{G}\left[\mathbf{x}(t), \mathbf{y}_a(t)\right] - \tau D_y \mathbf{G}\left[\mathbf{x}(t), \mathbf{y}_a(t)\right]\dot{\mathbf{y}}_a(t), \tag{2.78}$$

where D_y is the Jacobian operator with respect to $\mathbf{y}$.

Combining Equations 2.77 and 2.78 and replacing $\mathbf{y}_a$ by $\mathbf{y}(t)$, we obtain a new response system

$$\left(\mathbf{I} + \tau D_y \mathbf{G}(\mathbf{x}, \mathbf{y})\right) \dot{\mathbf{y}} = \mathbf{F}(\mathbf{y}) + \mathbf{G}(\mathbf{x}, \mathbf{y}), \qquad (2.79)$$

where $\mathbf{I}$ is the $n \times n$ identity matrix. For $\tau = 0$, the new response system completely synchronizes with the drive. For small detuning τ, the response allows an approximate time-shifted synchronization that lags or anticipates the drive with the time shift τ.

In fact, Equation 2.79 is a new, intentionally mismatched, response system that yields lag ($\tau > 0$), identical ($\tau = 0$), and anticipating ($\tau < 0$) synchronization without requiring an explicit time delay or memory in the coupling. Lag results if the response system dynamics are slower that the drive, whereas anticipation occurs if the response is faster.

Lag and anticipating synchronization can be detected by using the cross-correlation function C_{ij} between the times series of the oscillators i and j and the similarity function S_{ji}, defined, respectively, as (Rosenblum et al. 1997)

$$C_{ji}(\tau) = \frac{\langle [x_j(t) - \langle x_j \rangle][x_i(t + \tau) - \langle x_i \rangle] \rangle}{\sqrt{[x_j(t) - \langle x_j \rangle]^2 [x_i(t) - \langle x_i \rangle]^2}} \qquad (2.80)$$

and

$$S_{ji}^2(\tau) = \frac{\langle (x_j(t) - x_i(t + \tau))^2 \rangle}{\sqrt{\langle x_j(t)^2 \rangle \langle x_i(t)^2 \rangle}}, \qquad (2.81)$$

A higher maximum of the cross-correlation and a lower minimum of the similarity indicate a better synchronization.

When the coupled systems are in the state of lag or anticipating synchronization, the minimum of the similarity function vanishes at some value of time τ that is the time shift between the state vectors of the interacting systems. Obviously, if $\min_\tau S(\tau) = 0$ occurs at $\tau = 0$, the systems are in a complete synchronization state.

Figure 2.17 shows how the extrema of the two functions (cross-correlation and similarity) depend upon the coupling strength, for the case described before of three Rössler oscillators coupled unidirectionally in a ring.

One easily sees that $\max_\tau C(\tau)$ and $\min_\tau S(\tau)$ are negatively correlated. For small coupling ($\sigma < 0.048$) in the region of intermittent phase synchronization, $\max_\tau C(\tau)$ is very low, while $\min_\tau S(\tau)$ is very high. As the coupling increases from $\sigma = 0.048$ to $\sigma = 0.18$, imperfect phase synchronization becomes perfect thus resulting in a slowly increasing $\max_\tau C(\tau)$ and a slowly decreasing $\min_\tau S(\tau)$.

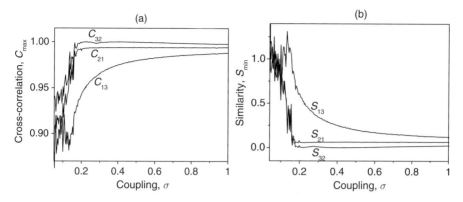

Figure 2.17 (a) Cross-correlation function and (b) similarity function versus the coupling strength, for each pair of oscillators.

In this regime, the phase difference ϑ fluctuates around its average value. As σ further increases, the amplitude of these fluctuations decreases leading to perfect phase synchronization.

On the route from phase to lag or anticipating synchronization, the oscillators adjust their amplitudes, as σ is further increased. Although complete synchronization cannot be achieved in this system because the oscillators are not identical, for a strong enough coupling the amplitudes become strongly correlated, with time lag or anticipation depending on the sign of the frequency mismatch. The time shift between the waveforms of the slave and master oscillators is always equal to $2\pi/(1/\omega_j - 1/\omega_i)$, where ω_j and ω_i are the natural frequencies of the slave and master oscillators.

The oscillators with closer natural frequencies synchronize better than the oscillators with larger mismatch. It should be noted that the best level of synchronization is achieved between oscillators 3 and 2, because oscillator 2 is equidistant from its neighbors. For a very strong coupling, the waveforms of oscillators 3 and 2 are completely identical but shifted in time, with either lag or anticipation depending on the sign of the frequency mismatch.

Changes in the type of synchronization are accompanied by and associated with a change in sign of one or more Lyapunov exponents. All nine Lyapunov exponents of the system are plotted in Figure 2.18 as a function of the coupling strength. One can see that the system of three coupled oscillators is hyperchaotic for any σ, because two or more Lyapunov exponents are always positive.

The two largest Lyapunov exponents λ_1 and λ_2 are always positive, while the four smallest, λ_{6-9}, are always negative, and the sign of the three remaining exponents, λ_{3-5}, depends on σ. Thus, these three exponents determine the type of synchronization.

Low-Dimensional Systems

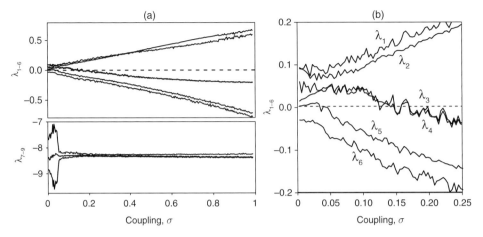

Figure 2.18 (a) Lyapunov exponents as a function of the coupling strength. (b) Enlarged part of panel (a), zooming on the behavior of the Lyapunov exponents for the weak coupling regime, associated with intermittent phase synchronization (for $\sigma > 0.05$) and phase synchronization (for $\sigma > 0.2$).

One can see from Figure 2.18 that intermittent phase synchronization takes place for $\sigma < 0.05$ when five Lyapunov exponents are positive ($\lambda_{1-5} > 0$) and four are negative ($\lambda_{6-9} < 0$). Phase synchronization starts at the point where λ_5 passes from zero to negative values. Finally, anticipating synchronization appears at $\sigma \approx 0.2$ when two more Lyapunov exponents ($\lambda_{3,4}$) cross zero.

2.7 Coherence Enhancement

We have already shown that the dynamics of a chaotic system can be regularized by the interaction with other systems to reach a synchronous state. Indeed, synchronization is an example of self-organization in nature (Camazine et al. 2001; Strogatz 2003), and it is usually assumed that interaction between oscillators enhances synchronization. However, this is not always the case.

In fact, an increased coupling between chaotic systems may result in unexpected behaviors, such as oscillation death (Bar-Eli 1985; Matthews and Strogatz 1990) and coherence enhancement (Bragard et al. 2007; Pisarchik and Jaimes-Reátegui 2015). While oscillation death reveals itself as complete disappearance of any oscillations assuming a stable steady state, the coherence enhancement, instead, implies the suppression of chaos to more regular, almost periodic, oscillations.

Chaos suppression in coupled chaotic oscillators was found in two cases, in the presence of asymmetry in coupling in bidirectionally coupled chaotic systems (Bragard et al. 2007) and in the presence of a small mismatch between parameters of unidirectionally coupled chaotic oscillators (Pisarchik and Jaimes-Reátegui 2015; García-Vellisca et al. 2016; Pisarchik et al. 2017). In the latter case, a chaotic master system forces a chaotic slave system towards almost regular oscillations with the same dominant frequency in the power spectrum. In terms of synchronization theory, this means that the oscillators are in a phase synchronization state, i.e., they develop a perfect phase-locking relation, although their amplitudes remain almost uncorrelated.

In order to demonstrate the effect of coherence enhancement, we consider a system of unidirectionally coupled chaotic oscillators:

$$\dot{\mathbf{x}}_j = \mathbf{F}(\mathbf{x}_j, \omega_j) + \sigma_{ji}(\mathbf{x}_i - \mathbf{x}_j).$$

Notice that here the oscillators are only distinct because of their natural frequencies ($\omega_j \neq \omega_i$). The dominant frequency Ω_j in the chaotic power spectrum of the uncoupled jth oscillator usually does not coincide with its natural frequency ($\Omega_j \neq \omega_j$). When the oscillators are unidirectionally coupled, the ith oscillator drives the jth oscillator. For sufficiently strong coupling, the master oscillator i entrains the dominant frequency Ω_j of the slave oscillator j, resulting in phase synchronization. The time-averaged difference between the oscillators' phases $\delta_{ji} = \langle \varphi_j - \varphi_i \rangle$ is negative if the frequency mismatch $\Delta_{ji} = \omega_j - \omega_i < 0$ and positive if $\Delta_{ji} > 0$. The phase of the slave oscillator is entrained by the master oscillator, with lag in the former case, and with anticipation in the latter.

A small frequency mismatch not only leads to phase synchronization, but it can also induce deterministic coherence resonance in the slave system. This surprising phenomenon resembles a "stabilization of chaos by chaos." In other words, a chaotic system under a chaotic drive may behave more regularly, almost periodically.

Coherence enhancement can be revealed by analysis of the time series, and by constructing bifurcation diagrams of peak values of the system variables and inter-peak intervals (IPI).[2]

For a quantitative description, some measures can be used. The first is the standard deviation of the peak amplitude of the IPIs (or ISIs), normalized to the mean value, referred to as *normalized standard deviation* (NSD); the smaller the NSD, the higher the coherence. A second measure is the relative width of the spectral component at the dominant frequency of the power spectrum; the narrower the peak, the higher the coherence. It should be mentioned that this measure can induce spurious results in systems with several distinguishable peaks with comparable heights but different widths placed at incommensurate frequencies, i.e., in systems with different time scales. In this case, it is not possible to decide univocally the peak to use as a measure of coherence, and therefore the criterion cannot be used.

---

[2] Or interspike intervals (ISI) in case of spiking systems, like neurons.

Coherence of a system with different time scales can also be measured by analyzing probability distribution of IPI (ISI). For example, in the case of bursting and spiking neurons, the interburst and interspike intervals are distributed over different time intervals. The half width at half height (HWHH) of each distribution will give information about coherence in the corresponding time scale.

2.7.1 Deterministic Coherence Resonance

Coherence enhancement can have a resonant character with respect to the parameter mismatch Δ_{ji}, as well as to the coupling strength σ of the chaotic oscillators.

Figure 2.19 illustrates the main features of the coherence enhancement in terms of the time series, the power spectra, and the corresponding attractors. First, the time series of the drive and response oscillators are shown in Figures 2.19(a) and 2.19(b) for the uncoupled and coupled case, respectively. When the oscillators are uncoupled (left-hand column), they behave chaotically with different dominant frequencies in their power spectra, namely $\Omega_1 = 1.07$ and $\Omega_2 = 1.16$ (see Figure 2.19(c)). When the oscillators are coupled ($\sigma = 0.2$), the dominant frequency of the slave oscillator Ω_2 is entrained by the master oscillator (see Figure 2.19(d)), resulting in phase synchronization. This is accompanied by an enhancement in the coherence of the dynamics of the slave oscillator, whose chaotic attractor shrinks, as can be seen by comparing Figure 2.19(e) with Figure 2.19(f).

The bifurcation diagrams of the amplitude (Figure 2.20(a)) and IPI (Figure 2.20(b)) with respect to the natural frequency ω_2 of the slave oscillator as a control parameter, display a strong resonant effect, yielding suppression of the amplitude variation at $\omega_2 \approx 1.1$. The coherent resonances, both in the amplitude and in time, are clearly seen in Figures 2.20(c) and 2.20(d), respectively.

In the space of the two control parameters (the natural frequency ω_2 of the slave oscillator and the coupling strength σ), both the NSDs of peak y_2 (Figure 2.21(a)) and the IPI (Figure 2.21(b)) feature an Arnold tongue structure centered at $\omega_2 = 1$. The dashed lines in the figures bound the phase-locking region, where the dominant frequency of the slave oscillator is entrained by the master oscillator, so that the coupled oscillators are in a phase synchronization state.

2.7.2 Stabilization of Periodic Orbits in a Ring of Coupled Chaotic Oscillators

When chaotic oscillators are unidirectionally coupled in a ring, they give rise to a periodic behavior for certain values of the coupling strength and parameter mismatch. This unexpected phenomenon of chaos stabilization can be observed in three Rössler oscillators coupled via their y variable (García-Vellisca et al. 2016). The time series, phase portraits, and power spectra in Figure 2.22

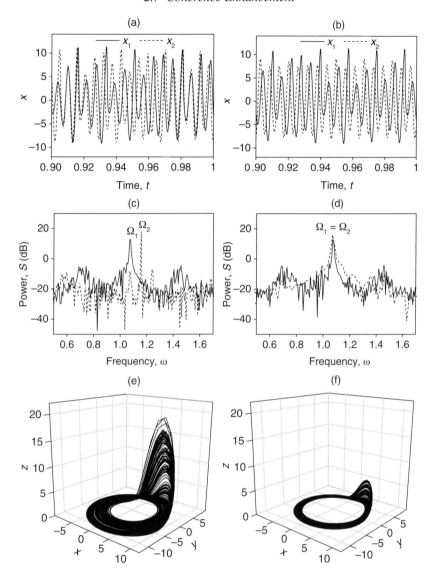

Figure 2.19 Coherence enhancement in two coupled chaotic Rössler oscillators with parameters $a = 0.2$, $b = 0.2$, $c = 5.7$, $\omega_1 = 1$, and $\omega_2 = 1.1$. (a,b) Time series of master (solid lines) and slave (dashed lines) oscillators, and (c,d) power spectra of master (solid) and slave (dashed) oscillators. In panels (a,c) $\sigma = 0$, in panels (b,d) $\sigma = 0.2$, (e) attractor of uncoupled slave oscillator and (f) attractor of coupled slave oscillator with coupling $\sigma = 0.2$. The coupling has the effect of shrinking the chaotic attractor of the slave oscillator, i.e., its dynamics become more regular.

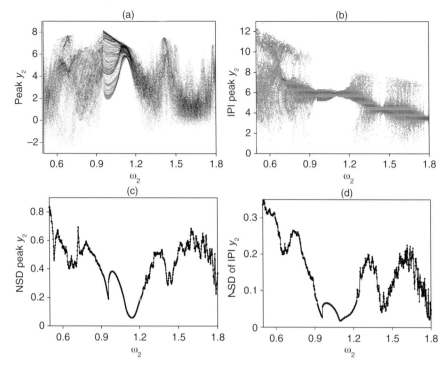

Figure 2.20 Coherent resonance behavior in coupled chaotic oscillators. (a) Peak values of y_2, (b) interpeak interval (IPI), (c) normalized standard deviation (NSD) of peak y_2, and (d) NSD of IPI as a function of natural frequency of slave oscillator ω_2, for $\sigma = 0.2$ and $\omega_1 = 1$. The strong suppression of the diagrams in (a) and (b) at $\omega_2 \approx 1.1$ results from resonant coherence enhancement.

illustrate the dynamics of the uncoupled and coupled oscillators. When uncoupled, the oscillators have different dominant frequencies (Figure 2.22(c)) in their chaotic power spectra. As the coupling strength is increased, they adjust their dominant frequencies (Figure 2.22(d)) and oscillate periodically in time (Figure 2.22(b)).

The stability analysis shows that for large mismatches ($\Delta > 0.1$) and intermediate coupling strengths ($0.1 < \sigma < 0.4$) the largest Lyapunov exponent takes a nonpositive value (Figure 2.23), and therefore the system exhibits periodic dynamics in that range.

2.8 Generalized Synchronization

In the previous sections, we have reviewed synchronization phenomena in coupled nonidentical chaotic oscillators. In general, when the interacting systems are essentially different (such as, for instance, if they have different phase-space dimensions), synchronization is not so trivial.

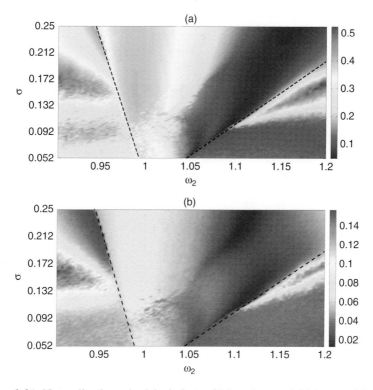

Figure 2.21 Normalized standard deviations of (a) peak y_2 and (b) interpeak intervals in the (ω_2, σ)-parameter space. The dashed lines bound the Arnold tongues within which the oscillators are in a phase synchronization state. The dark spots on the right-hand side inside the Arnold tongues indicate the regions where higher coherence is featured.

In these cases, the emerging state is what is called *generalized synchronization*, a situation where the trajectories of the coupled systems have a certain functional dependence on each other. This type of synchronization was first introduced for unidirectionally coupled nonidentical oscillators (Rulkov et al. 1995; Kocarev and Parlitz 1996; Pyragas 1996; Hramov and Koronovskii 2005; Poria 2007), as a state where the trajectory of the slave system is in a given functional relationship with that of the master. Later on, the conditions were also formulated for bidirectionally coupled systems (Khan and Poria 2012; Islam et al. 2013).

2.8.1 Generalized Synchronization in Unidirectionally Coupled Systems

In order to define generalized synchronization for the unidirectional coupling scheme, let us consider the following coupled system:

$$\begin{aligned} \dot{\mathbf{x}} &= \mathbf{F}(\mathbf{x}), \\ \dot{\mathbf{y}} &= \mathbf{G}(\mathbf{y}, \mathbf{H}(\mathbf{x})). \end{aligned} \qquad (2.82)$$

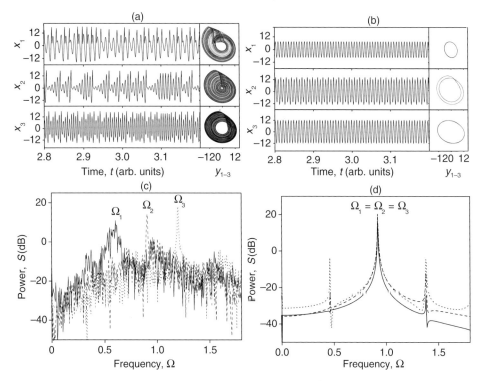

Figure 2.22 Stabilization of periodic orbits in three coupled chaotic Rössler oscillators. (a,b) Time series and phase portraits and (c,d) power spectra of (a,c) uncoupled oscillators and (b,d) coupled oscillators with coupling strength $\sigma = 0.33$. Other parameters are $\omega_2 = 0.9$ and $\Delta_{21} = \Delta_{32} = 0.2$. The parameters of the Rössler oscillators are $a = 0.165$, $b = 0.2$, and $c = 10$.

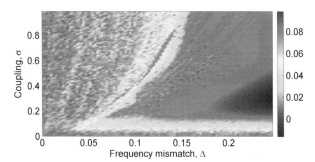

Figure 2.23 Largest Lyapunov exponent of three ring-coupled oscillators in (Δ, σ)-parameter space. The dark spot on the right-hand side (nonpositive exponent) is the region where all oscillators are in a periodic regime.

In this system, $\mathbf{x} \in \mathbb{R}^m$, $\mathbf{y} \in \mathbb{R}^n$, and $\mathbf{H}$ is a vector field that maps $\mathbb{R}^m$ into $\mathbb{R}^k$. It is said that the system displays generalized synchronization if there exists a transformation $\mathbf{M} : \mathbb{R}^m \to \mathbb{R}^n$ that defines a manifold $\mathbf{y} = \mathbf{M}(\mathbf{x})$, and a subset $B \subseteq \mathbb{R}^m \times \mathbb{R}^n$ such that all trajectories of the response system starting from the basin B converge to the manifold as time tends to infinity.

If $\mathbf{H}$ is the identity transformation, then generalized synchronization coincides with identical synchronization.

Several methods have been proposed for detection of a generalized synchronization state in unidirectionally coupled chaotic oscillators. The most popular are a technique based on the calculation of conditional Lyapunov exponents (Pecora and Carroll 1990; Pyragas 1997), the auxiliary system method (Abarbanel et al. 1996) and the method of nonlinear interdependence. We now briefly describe each of them.

Conditional Lyapunov Exponents

Generalized synchronization can be analyzed using conditional Lyapunov exponents (Pecora and Carroll 1991; Pyragas 1996) described in Section 2.4.3 in the context of synchronization of unidirectionally coupled oscillators.

The stability of an unidirectionally coupled chaotic system is characterized by $N = m + n$ Lyapunov exponents $\lambda_1 \geq \lambda_2 \geq \ldots \geq \lambda_{m+n}$. Since the dynamics of the drive oscillator are independent of the slave oscillator, the Lyapunov exponent spectrum can be separated into two parts, namely, the exponents $\lambda_1^d \geq \ldots \geq \lambda_m^d$ relative to the drive oscillator, and the exponents $\lambda_1^r \geq .. \geq \lambda_n^r$ relative to the response oscillator. The word "conditional" here means that those exponents are conditioned to the specific dynamics of the master.

A necessary condition for the existence of generalized synchronization is therefore that the largest conditional Lyapunov exponent is not positive, i.e., $\lambda_1^r \leqslant 0$.

Auxiliary System Method

The most efficient technique to detect generalized synchronization is the auxiliary system method (Abarbanel et al. 1996). The idea of this method consists in the creation of an auxiliary system $\mathbf{v}(t)$ identical to the response system $\mathbf{y}(t)$, but with different initial conditions, i.e., $\mathbf{v}(t_0) \neq \mathbf{y}(t_0)$, although both $\mathbf{v}(t_0)$ and $\mathbf{y}(t_0)$ must belong to the same basin of attraction in case of multistability.

If generalized synchronization exists in the unidirectionally coupled chaotic systems, the trajectories $\mathbf{y}(t)$ and $\mathbf{v}(t)$ converge and become equivalent after a transient because $\mathbf{y}(t) = \mathbf{f}(\mathbf{x}(t))$ and $\mathbf{v}(t) = \mathbf{f}(\mathbf{x}(t))$. Thus, the coincidence of the state vectors of the response and the auxiliary systems $\mathbf{y}(t) \equiv \mathbf{v}(t)$ is considered as a criterion of the presence of generalized synchronization. Note that, strictly speaking, the criterion is only a test of sufficience. In fact, $\mathbf{v}(t_0)$ could be outside the basin B, and therefore not converge at all to $\mathbf{y}(t)$. However, in practice, the test is very efficient if one chooses $\mathbf{v}(t_0)$ to be sufficiently close to $\mathbf{y}(t_0)$.

Nonlinear Interdependences

A further method is that of nonlinear interdependencies, which is very convenient in real-time analysis of time series, especially for noisy data or data coming from systems where the dynamical equations are unknown (Quian Quiroga et al. 2000). The basic idea is illustrated in Figure 2.24, where these measures are exemplified for the characterization of synchronization between systems with different struc-ture of attractors, as, for example, the Rössler and the Lorenz oscillators. Similarly to the nearest neighbor method, the size of the neighborhood in one of the systems is compared with the size of its mapping in the other system.

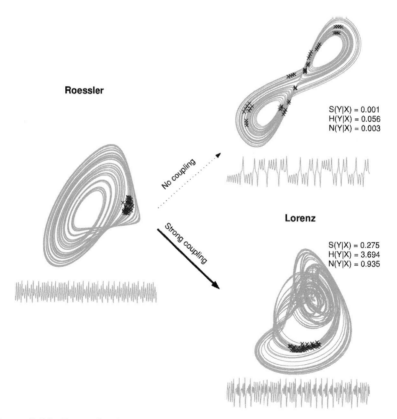

Figure 2.24 Generalized synchronization of the Rössler (**x**) and Lorenz (**y**) systems (reprinted with permission from Quian Quiroga et al. 2002). The corresponding time series is shown below each attractor.

The method works as follows. Suppose we have to coupled n-dimensional systems $\mathbf{u}$ and $\mathbf{v}$. Also let

$$x = f(\mathbf{u})$$

and

$$y = f(\mathbf{v}),$$

where f is a smooth scalar field $\mathbb{R}^n \to \mathbb{R}$. Now assume we have two simultaneously measured discrete time series, $x(t)$ and $y(t)$. Using the method of delay coordinates, given an embedding dimension m and a time lag τ, at any moment in time t, we create the two m-dimensional delay vectors

$$\mathbf{x}(t) = (x(t - (m-1)\tau), x(t - (m-2)\tau), \ldots, x(t - \tau), x(t))$$

and

$$\mathbf{y}(t) = (y(t - (m-1)\tau), y(t - (m-2)\tau), \ldots, y(t - \tau), y(t)).$$

A correct choice of the embedding parameters m and τ is very important for the phase-space reconstruction, and there is a large literature on the optimal choice of the embedding parameters (see, for example, Takens 1981 and references therein).

Since the time series $x(t)$ and $y(t)$ are discrete, the number of vectors $\mathbf{x}(t)$ and $\mathbf{y}(t)$ that can be created is finite. Then, for any given $\mathbf{x}(t)$, using a standard Euclidean norm in $\mathbb{R}^m$, consider the k $\mathbf{x}$-delay vectors closest to it, $\mathbf{x}(t_1), \mathbf{x}(t_2), \ldots, \mathbf{x}(t_k)$. Similarly, let $\mathbf{y}(t_1'), \mathbf{y}(t_2'), \ldots, \mathbf{y}(t_k')$ be the k $\mathbf{y}$-delay vectors closest to $\mathbf{y}(t)$.

Then, compute the mean square Euclidean distance of $\mathbf{x}(t)$ to its k nearest neighbors

$$R_t(\mathbf{x}) = \frac{1}{k} \sum_{j=1}^{k} \left(\mathbf{x}(t) - \mathbf{x}(t_j) \right)^2$$

and the $\mathbf{y}$-conditioned mean square Euclidean distance

$$R_t(\mathbf{x}|\mathbf{y}) = \frac{1}{k} \sum_{j=1}^{k} \left(\mathbf{x}(t) - \mathbf{x}(t_j') \right)^2,$$

obtained by replacing the nearest neighbors of $\mathbf{x}(t)$ with the $\mathbf{x}$-delay vectors of the same times as the nearest neighbors of $\mathbf{y}(t)$. These two quantities can be compared in different ways to quantify interdependence between x and y and, ultimately, between $\mathbf{u}$ and $\mathbf{v}$.

A very common choice is to compute

$$S(\mathbf{x}|\mathbf{y}) = \left\langle \frac{R_t(\mathbf{x})}{R_t(\mathbf{x}|\mathbf{y})} \right\rangle_t,$$

where the average is taken over the whole discretized time series. Low values of S indicate the absence of interdependence, while values close to 1 suggest that the systems are in a state of generalized synchronization.

2.8.2 Generalized Synchronization in Bidirectionally Coupled Systems

Generalized synchronization in bidirectionally coupled chaotic systems $\mathbf{x} \in \mathbb{R}^n$ and $\mathbf{y} \in \mathbb{R}^n$ is defined as follows (Islam et al. 2013). Two chaotic systems:

$$\begin{aligned} \dot{\mathbf{x}} &= \mathbf{f}(\mathbf{x}, \mathbf{y}), \\ \dot{\mathbf{y}} &= \mathbf{g}(\mathbf{x}, \mathbf{y}), \end{aligned} \tag{2.83}$$

are said to be in a state of generalized synchronization if for a constant invertible matrix

$$\mathbf{D} = \begin{pmatrix} d_{11} & d_{12} & \cdots & d_{1n} \\ d_{21} & d_{22} & \cdots & d_{2n} \\ \vdots & \vdots & \ddots & \vdots \\ d_{n1} & d_{n2} & \cdots & d_{nn} \end{pmatrix}, \tag{2.84}$$

the condition

$$\lim_{t \to \infty} |\mathbf{x} - \mathbf{D}\mathbf{y}| = 0, \tag{2.85}$$

is satisfied.

There are at least two methods to achieve generalized synchronization. In the first method, the drive and response systems are considered as

$$\begin{aligned} \dot{\mathbf{x}} &= A\mathbf{x} + \mathbf{f}(\mathbf{x}, t) + \mathbf{V}_1(\mathbf{y}, t) && \text{Drive} \\ \dot{\mathbf{y}} &= A\mathbf{y} + \mathbf{g}(\mathbf{y}, t) + \mathbf{V}_2(\mathbf{x}, \mathbf{y}, t), && \text{Response} \end{aligned} \tag{2.86}$$

with coupling functions $\mathbf{V}_1$ and $\mathbf{V}_2$ given as

$$\begin{aligned} \mathbf{V}_1(\mathbf{y}, t) &= D\mathbf{g}(\mathbf{y}) + (DA - AD)\mathbf{y}, \\ \mathbf{V}_2(\mathbf{x}, \mathbf{y}, t) &= D^{-1}\mathbf{f}(\mathbf{x}) + D^{-1}BK(\mathbf{x} - D\mathbf{y}), \end{aligned} \tag{2.87}$$

where A is a $n \times n$ matrix. Defining the error as $\mathbf{e} = \mathbf{x} - D\mathbf{y}$, from Equations 2.86 and 2.87 we get

$$\dot{\mathbf{e}} = (A - BK)\mathbf{e}, \tag{2.88}$$

where K is a $1 \times n$ feedback matrix and B is a $n \times 1$ suitable matrix (Khan and Poria 2012). If all eigenvalues of the matrix $(A - BK)$ have negative real parts, then the condition for generalized synchronization Equation 2.85 is fulfilled for the system in Equation 2.86 with $\mathbf{V}_1$ and $\mathbf{V}_2$ given by Equation 2.87.

In another technique, the drive and response systems are considered as follows

$$\dot{\mathbf{x}} = A\mathbf{x} + \mathbf{f}(\mathbf{x}, t) + \mathbf{h}_1(\mathbf{x}, \mathbf{y}), \qquad \text{Drive}$$
$$\dot{\mathbf{y}} = A\mathbf{x} + \mathbf{g}(\mathbf{y}, t) + \mathbf{h}_2(\mathbf{x}, \mathbf{y}) + \mathbf{U}, \qquad \text{Response} \tag{2.89}$$

where $\mathbf{h}_1$ and $\mathbf{h}_2$ are coupling functions and $\mathbf{U}$ is a control function. Taking the error as $\mathbf{e} = \mathbf{x} - D\mathbf{y}$, the error dynamics of the system in Equation 2.89 become

$$\dot{\mathbf{e}} = A\mathbf{e}, \tag{2.90}$$

provided by the control in the form:

$$\mathbf{U} = D^{-1}\left[\mathbf{f}(\mathbf{x}, t) + \mathbf{h}_1(\mathbf{x}, \mathbf{y})\right] - \mathbf{g}(\mathbf{y}, t) - \mathbf{h}_2(\mathbf{x}, t) - A\mathbf{y} + D^{-1}AD\mathbf{y}. \tag{2.91}$$

If the real parts of all eigenvalues of A are negative, then the system Equation 2.90 is asymptotically stable at the origin, and therefore the coupled system in Equation 2.89 is in generalized synchronization.

2.9 A Unifying Mathematical Framework for Synchronization

So far, we have discussed several types of synchronization occurring when dynamical systems are coupled. An interesting issue is whether a unifying mathematical framework exists for synchronization of coupled dynamical systems, i.e., whether one can formulate a unique definition of synchronization that encompasses the various types of synchronous motions that are observed in nature, so that different kinds of synchronization might be captured by a single formalism.

The problem was first approached with the assumption that a system is divisible into two subsystems, for whose properties one can define functions that are mappings from the space of trajectories (plus time) to some Cartesian space (Boccaletti et al. 2001).

Mathematically, this looks as follows. The total $(m + n)$-dimensional system is given by $z = [x, y]$, with x being a m-dimensional system, and y a n-dimensional one. The subsystems form trajectories $\varphi_x(z_0)$ and $\varphi_y(z_0)$ starting from a generic initial condition z_0, and these trajectories are mapped to a new d-dimensional space by two functions (or properties), g_x and g_y. Global synchronization requires a function $h(g_x, g_y)$ on these trajectories with either $|h| = 0$ or $|h| \to 0$ when $t \to \infty$.

In fact, one can even simplify and generalize this definition of synchronization to a more condensed and concrete form (Boccaletti et al. 2001), starting from the same assumptions. Typically, when one states that a system is synchronized to another, one means that a certain event in a given subspace always happens when some other particular event in the other subspace occurs, i.e., one can predict the value or state of one subsystem from that of the other. Now, one can identify events with points in the state space of the subsystems, and state that synchronization happens when there is a local function from one subspace to another such that a particular point $\tilde{x}$ of one subspace is mapped, uniquely, to one point $\tilde{y}$ of the other subspace.

However, in order to be realistic, one has to consider that, when searching for synchronization in data or numerical calculations, one never has points that fall *exactly* on a given $\tilde{x}$ or $\tilde{y}$. Rather, the closer $x(t)$ is to $\tilde{x}$, the closer $y(t)$ is to $\tilde{y}$, a statement that is captured rigorously by a *continuous* local function. In other words, the trajectories of $x(t)$ close to $\tilde{x}$ are mapped to points close to $\tilde{y}$ by a function that is continuous at the point $(\tilde{x}, \tilde{y})$.

The mathematical formalism has been fully developed (Boccaletti et al. 2001), and the concepts of a local synchronization function and local synchronization points are well defined: a set of local functions f_i are said to be local synchronization functions when they represent the mapping of the trajectories in the proximity of the corresponding synchronization points $(\tilde{x}_i, \tilde{y}_i)$.

Using this formalism, one can prove a theorem that states the conditions for which *global* synchronization emerges: given a set of synchronization points $(\tilde{x}_i, \tilde{y}_i)$, the associated set of synchronization functions must provide a continuity covering of the attractor in the subsystem x.

Furthermore, the various types of synchronization (identical, lag, phase, anticipating, and generalized) are actually all specific cases of global synchronization phenomena. Proving these results is out of the scope of this book, and we refer the interested reader to (Boccaletti et al. 2001), where the demonstrations can be found.

Later on, Pastur and collaborators (2004) introduced a technique that, based on the same mathematical formalism, detects and quantifies local functional dependencies between coupled chaotic systems. The fraction of locally synchronized configurations or *synchronization point percentage* (SPP) can be used as a measure for quantifying synchronization of a pair of arbitrary signals within an arbitrary synchronization state. SPP is a quantity ranging from 0 (no points in the subspaces admit local synchronization functions) to 1 (for full synchronization), and can also quantify the number of dynamical configurations where a local prediction task is possible, even in the absence of global synchronization features. When applied to a pair of interacting Rössler oscillators, SPP shows that, during phase synchronization, SPP already deviates significantly from 0.

3

Multistable Systems, Coupled Neurons, and Applications

Multistability is a phenomenon occurring in dissipative systems when several stable attractors coexist for a given set of the system parameters. This phenomenon has been observed in many fields of science, including electronics (Maurer and Libchaber 1980), optics (Brun et al. 1985; Ruiz-Oliveras and Pisarchik 2009), mechanics (Thompson and Stewart 1986), and biology (Foss et al. 1996). The mechanisms responsible for multistability can be many and diverse, including homoclinic tangencies (in weakly dissipative systems) and delayed feedback.

In spite of the possible differences in the origin of multistability, the behaviors of multistable systems share several similarities. All multistable systems are characterized by an extremely high sensitivity to initial conditions, so that even very tiny perturbations can cause a significant change in the final attractor state. Furthermore, their qualitative behavior often changes drastically under just very small variations of their parameters. Finally, the intervals of coexistence of attractors can be so small that a slight perturbation in a control parameter can cause a rapid change in the number of coexisting attractors, giving rise to very complex dynamics.

Synchronization of multistable chaotic systems is particularly relevant to sociology and brain research, where it helps scientists acquire a better understanding of the mechanisms underlying the formation of social groups, the sudden emergence of new ideas, the adoption of innovations, the mechanisms of decision making, and the spread of habits, fashions, leading opinions, and panic. Considering individuals or groups of people as dynamical units, one could say that coordination in their behavior is a type of synchronization. For example, in a crowd, people lose their individuality and act synchronously as a common system. This occurs very frequently when people are affected by a common external force, for instance when they look at or listen to the same object, be it a political leader, a movie, a football game, or a rock concert.

Psychologists and sociologists have long been studying this phenomenon. The first classic treatment of crowds was offered more than 100 years ago by Gustave LeBon (1896), who interpreted the mob behavior during the French Revolution as an irrational reversion to wild animal emotions. Later, a similar view was expressed by Freud (1922). However, only relatively recently has crowd behavior been treated as a collective phenomenon (Turner and Killia 1993; McPhail 1991), as experimental studies showed that it results from self-organized processes based on local interactions among individuals (Moussaïd et al. 2009). Subsequently, the definition of crowd was extended also to participants who are not assembled in one single physical place but dispersed over a large area (*diffuse crowds*) (Turner and Killia 1993). For instance, when watching a football game in your home, you will feel similar emotions and act synchronously with other people who watch the same game in other places.

Thus, mathematicians and physicists tried to explain the collective behavior of individuals in terms of nonlinear dynamics, developing a series of simple models (Coscia and Canavesio 2008; Castellano et al. 2009; Bellomo et al. 2009; Bruno et al. 2011; Dostalkova and Splnka 2010). At the same time, some psychologists and sociologists applied physical approaches to explain complex behaviors of the brain in terms of synchronization, multistability, and complex network theory (see, for example, Cramer et al. 2010 and references therein). In fact, modern trends in physiology, psychology, and cognitive neuroscience suggest that chaos is the type of motion that provides the self-organization needed for the functioning of the brain (Basa 1990; Bob 2007; Kitzbichler et al. 2009). Numerous experiments on neuronal activity, including the analysis of recorded electroencephalograms, prove that neuron dynamics exhibit many features inherent to chaos and confirm a possible role of chaotic transitions in the processing of dissociated memory. More recently, Kitzbichler et al. (2009) argued that the brain operates at a critical point between organization and chaos, a state previously described as self-organized criticality. Moreover, many researchers find that, in addition to chaos, brain dynamics display multistability (Atteneave 1971), which can be revealed by the study of the phase transitions that occur when changing operating conditions (preliminary knowledge and experience) or applying an external influence (Jirsa et al. 1994; Wallenstein et al. 1995).

Multistability studies of visual and auditory perception were among the first to provide mechanisms for information processing in biological systems (Kruse and Stadler 1995; Warren 1999). Brain multistability was also indicated as potentially responsible for memory and pattern recognition (Hertz et al. 1991; Canavier et al. 1993): the coexisting attractors define different states of the brain corresponding to particular objects of perception that can be selected by the input to the neural network (Kim et al. 1997). This suggests that an adequate mathematical model to describe brain behavior should be based on a dynamical system that exhibits the coexistence of chaotic attractors.

Nonetheless, the prediction of synchronization in multistable systems is still a largely debatable question even in relatively simple systems such as iterative maps, because, as mentioned above, multistable systems are extremely sensitive to perturbations due to their complexly interwoven basins of attraction.

Thus, most studies of synchronization deal with monostable systems that exhibit a single attractor when uncoupled, but are multistable when they interact (Postnov et al. 1999). For example, Astakhov and his colleagues (2001; 2003) observed multistability in two mutually coupled Hénon maps, while Yanchuk and Kapitaniak (2003) found the coexistence of chaotic attractors in two identical bidirectionally coupled Rössler systems, and Guan et al. (2005) unveiled similar results for unidirectionally coupled Lorenz and Rössler oscillators.

In this chapter, we describe the main methodology for studying synchronization of multistable oscillators. First, we demonstrate the high complexity of the basins of attraction of two coexisting chaotic attractors in a solitary oscillator, and the even higher complexity of the basins when the two oscillators interact with each other. Then, we show how synchronization arises in unidirectionally and bidirectionally coupled multistable oscillators, with the help of some analytical tools, and with reference to experiments in coupled semiconductor lasers. The chapter ends with a discussion of synchronization in coupled neurons, and applications of synchronization for secure communication schemes.

3.1 Unidirectionally Coupled Multistable Systems

As in the case of generalized synchronization, we consider here a dynamical system divisible into two subsystems, namely an m-dimensional drive and an n-dimensional response. Furthermore, we assume that each uncoupled subsystem exhibits the coexistence of multiple attractors $A = \{A_1, \ldots, A_p\}$ and $B = \{B_1, \ldots, B_k\}$, for the drive and response systems, respectively.

Since the coupling is unidirectional, the response system has no influence on the drive, and thus the drive system always stays in one of its attractors, say A_i, defined by the initial condition $\mathbf{x_0} \in \mathbb{R}^m$. Conversely, the structure of the attractors in the response system can drastically change due to the coupling.

3.1.1 Synchronization States

For the sake of simplicity, we start by supposing that the drive and response subsystems are identical. Depending on the coupling strength, different stages of attractor metamorphosis are possible:

(i) For a very weak (approaching zero) coupling, the drive has almost no influence on the response system, and therefore the structure of attractors does not change, i.e., the attractors of the response system are identical to those of the drive system ($B = A$) and only determined by the initial conditions.

(ii) For weak couplings, the chaotic driving acts in a way similar to an external noise, and induces random switches between coexisting states in the response system. This state is a precursor of synchronization.

(iii) A further increase in the coupling strength results in intermittent phase synchronization. The phases synchronize only within time windows where the response system stays in an attractor B_j that corresponds to the attractor A_i of the drive system.

(iv) Intermediate coupling values may result in a homeomorphism (Afraimovich et al. 1986) that gives rise to generalized synchronization. This occurs if and only if the following conditions are true: (Kocarev and Parlitz 1996)

(a) There exists an attracting manifold $M = \{(\mathbf{x}, \mathbf{y}) \in \mathbb{R}^{2n}\}$ given by some function $\mathbf{H} : \mathbb{R}^n \to \mathbb{R}^n$ such that $\mathbf{y} = \mathbf{H}(\mathbf{x})$.

(b) M must possess an open basin $D \supset M$ such that $\lim_{t\to\infty} |\mathbf{y}(t) - \mathbf{H}(\mathbf{x}(t))| = 0$ for all initial conditions in D.

Note that the definition above also includes the case of subharmonic entrainment of periodic or chaotic oscillations with the dominant frequency of the drive system attractor f_{A_i}. For instance, calling the average periods of the oscillation of the drive and response systems T_{A_i} and T_{B_j}, one can have an entrainment with ratio $T_{A_i} : T_{B_j} = 1 : q$.

(v) Eventually, for very strong couplings, the coupling dominates over the evolution of the variables of the response system, leading to the annihilation of all attractors different from the attractor A_i. As soon as the coupling exceeds a threshold value, complete synchronization occurs, and all trajectories of the response system converge to the same trajectory of the driving one. The number of coexisting attractors in the response system therefore reduces to a single one, and a full entrainment takes place.

It is worth mentioning that synchronization of multistable systems is closely related to the problem of controlling multistability (Pisarchik and Feudel 2014), since in unidirectionally coupled systems the drive can be considered as a controller that pushes the trajectory of the response system towards a desired one.

3.1.2 An Example

As an illustration of synchronization in a chaotic multistable system, we discuss the case of two unidirectionally coupled identical piecewise-linear Rössler-like oscillators: (García-López et al. 2005; Pisarchik and Jaimes-Reátegui 2005a; Pisarchik et al. 2006, 2008a,b)

$$\dot{x}_1 = -\alpha x_1 - \beta x_2 - x_3, \quad \dot{y}_1 = -\alpha y_1 - \beta \left[y_2 + \sigma \left(x_2 - y_2 \right) \right] - y_3,$$
$$\dot{x}_2 = x_1 + \gamma x_2, \qquad\qquad \dot{y}_2 = y_1 + \gamma \left[y_2 + \sigma \left(x_2 - y_2 \right) \right], \qquad (3.1)$$
$$\dot{x}_3 = g\left(x_1\right) - x_3, \qquad\quad \dot{y}_3 = g\left(y_1\right) - y_3,$$

where

$$g\left(z\right) = \begin{cases} 0 & \forall z \leqslant 3 \\ \mu\left(z - 3\right) & \forall z > 3 \end{cases} \qquad (3.2)$$

is a piecewise-linear function.

In spite of its apparent simplicity, the system in Equation 3.1 exhibits very rich dynamics, including multistability, in contrast to the classical (smooth) Rössler oscillator, which does not display coexistence of attractors.

Coexistence of two chaotic attractors is observed for $\alpha = 0.05$, $\beta = 0.5$, $\gamma = 0.2925$, and $\mu = 15$. The trajectories of these coexisting states occupy different regions of the phase space, as seen from Figure 3.1.

The power spectra of these two attractors have different dominant frequencies, as can be seen from Figure 3.2.

Linear Stability Analysis

It should be noticed that the same dynamical system may exhibit attractors of different types (fixed point, periodic orbit, or chaotic) depending on the choice of parameters, as we illustrate below.

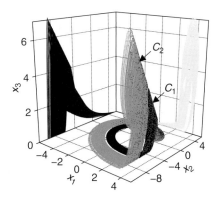

Figure 3.1 Chaotic trajectories C_1 and C_2 (originated from two different initial conditions) of the piecewise Rössler oscillator (Equation 3.1), and their projections on (x_1, x_3) and (x_2, x_3) planes.

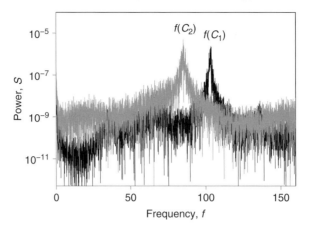

Figure 3.2 Power spectra of the x_2 variable of Equation 3.1, for the two coexisting chaotic regimes C_1 (black) and C_2 (grey). The dominant frequencies are $f(C_2)$ and $f(C_1)$.

Consider only one of the oscillators in Equation 3.1. It is not difficult to show that this system has two equilibrium points, $\mathbf{x}^*$ and $\mathbf{x}^{**}$, which can be found by solving the system of equations

$$-\alpha x_1 - \beta x_2 - x_3 = 0$$
$$x_1 + \gamma x_2 \quad\;\;= 0 \tag{3.3}$$
$$-x_3 = -g\,(x_1)\,.$$

These fixed points are

$$\mathbf{x}^* = (0, 0, 0)^{\mathrm{T}} \qquad\qquad\qquad \forall x_1 \leqslant 3,$$
$$\mathbf{x}^{**} = \left(3a, -\frac{3\mu\,(a-1)}{\beta - \alpha\gamma}, 3\mu\,(a-1)\right)^{\mathrm{T}} \forall x_1 > 3, \tag{3.4}$$

where $a = \gamma\mu/(\alpha\gamma - \beta + \gamma\mu)$.

A linear approximation of the system yields the Jacobian matrices

$$J_{\mathrm{F}}(\mathbf{x}^*) = \begin{pmatrix} -\alpha & -\beta & -1 \\ 1 & \gamma & 0 \\ 0 & 0 & -1 \end{pmatrix} \tag{3.5}$$

and

$$J_{\mathrm{F}}(\mathbf{x}^{**}) = \begin{pmatrix} -\alpha & -\beta & -1 \\ 1 & \gamma & 0 \\ \mu & 0 & -1 \end{pmatrix}. \tag{3.6}$$

The point $\mathbf{x}^*$ has two complex conjugated eigenvalues $\lambda_{1,2}^* = \rho^* \pm i\omega^*$ and one real eigenvalue $\lambda_3^* = -1$. Simple calculations show that, for $\alpha = 0.05$,

$\beta = 0.5$, and $\mu = 15$, the real part ρ^* is positive for any $\gamma > 0.05$. There-fore, this equilibrium point is an *unstable spiral*. Since one of the eigenvalues (λ_3^*) is real and negative, this fixed point has a stable manifold along which the tra-jectory approaches it without oscillations. The positive real parts of the complex conjugated eigenvalues indicate that the trajectory leaves this point along an unsta-ble manifold by making an untwist spiral with eigenfrequency ω. Conversely, for $\gamma \leqslant 0.05$, the real part is negative, and therefore the point is a *stable focus*.

The point $\mathbf{x}^{**}$ has one real eigenvalue λ_1^{**} and two complex conjugated eigen-values $\lambda_{2,3}^{**} = \rho^{**} \pm i\omega^{**}$. For our parameters, when $\gamma > 0.03$ the real eigenvalue is positive, and therefore the corresponding manifold is unstable. Note that even though the attractor is unstable, the fact that $\rho^{**} \approx -0.5$ indicates the existence of a stable manifold. The occurrence of all these conditions implies that $\mathbf{x}^{**}$ is an *unstable focus*: the trajectory approaches this fixed point by a twist spiral and leaves it without oscillations.

Given the structure of their corresponding Jacobian eigenvalues, $\mathbf{x}^*$ and $\mathbf{x}^{**}$ are both saddle points.

Bifurcation Diagrams

Figure 3.3 shows the bifurcation diagram of the local minima of the variable x_2 of the system in Equation 3.1 as a function of the control parameter γ. The dia-gram is calculated by taking many different initial conditions, and illustrates the superposition of the different coexisting attractors for each parameter.

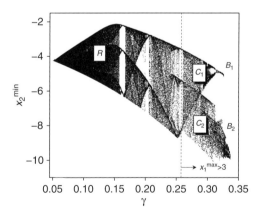

Figure 3.3 Bifurcation diagram of the local minima of the variable x_2 in Equa-tion 3.1, with γ taken as a control parameter. The dashed line indicates the parameter value where the second fixed point is born. R is the region of Rössler-type chaos and C_1 and C_2 are the coexisting chaotic attractors belonging to the branches B_1 and B_2, and found for distinct initial conditions.

For small γ, $x_1 < 3$ and therefore the only one fixed point is $\mathbf{x}^*$. In this range, the dynamics of the piecewise system are similar to those of the classical Rössler oscillator (Rössler 1976). They exhibit a period-one limit cycle for around the focus point $\gamma < 0.05$ and then, as γ is increased, undertake a period-doubling route to chaos.

As soon as the maximum value of the x_1 component of the chaotic trajectory exceeds 3, the second fixed point $\mathbf{x}^{**}$ appears, giving rise to a different chaotic attractor that coexists with the Rössler-type chaos C_1 for $\gamma > 0.25$. When $x_1 \leqslant 3$, the dynamics of the chaotic motion C_1 are characterized by a single fixed point at the origin. Conversely, when $x_1 > 3$, the behavior of the attractor C_2 is determined by both saddle points.

For large γ, the diagram contains two coexisting branches, B_1 and B_2, obtained for different initial conditions. Since the branch B_1 is a logical extension of the bifurcation diagram for $\gamma < 0.25$, chaos C_1 in this branch has dynamical features similar to those of the classical Rössler chaos. Another branch B_2, born at the critical point $\gamma \approx 0.25$ due to the collision of the chaotic trajectory with the saddle, gives rise to the homoclinic chaos first described by the Russian mathematician Shilnikov (1965). For $0.25 \lesssim \gamma \lesssim 0.29$ the two attractors, C_1 and C_2, are intermingled. Then, for $\gamma \gtrsim 0.29$, their trajectories separate and occupy different regions of the phase space, as shown in Figure 3.1. For even larger γ, stable periodic homoclinic orbits arise connecting the fixed point $\mathbf{x}^{**}$ with itself (Pisarchik and Jaimes-Reátegui 2005a).

Basins of Attraction

The basins of attraction of C_1 and C_2 have a very complex structure in phase space. Figure 3.4 shows the Poincaré section of the two coexisting chaotic attractors for $x_3 = 0.0092$ and the corresponding section of their basins of attraction.

One can see that the basin of attraction of C_2 is much larger than that of C_1. The basins form two pairs of nested spirals with their foci on the fixed points $\mathbf{x}^*$ and $\mathbf{x}^{**}$. One can also see that the attractors C_1 and C_2 almost touch the basin boundaries. This makes it possible to control the attractors by applying a small perturbation (or a small noise) that induces intermittent switches between these coexisting states.

Synchronization States

Consider now the full system of two oscillators described by Equation 3.1. Also, let us assume that, starting from different initial conditions, the uncoupled drive $\mathbf{x}$ and response $\mathbf{y}$ enter two distinct attractors, C_1 and C_2, respectively, which are never left. Finally, let the attractors be characterized by different amplitudes of the variables x_2 and y_2.

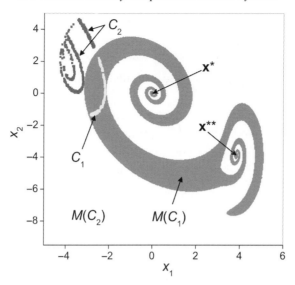

Figure 3.4 Poincaré section of the two coexisting attractors C_1 (white) and C_2 (black), with their basins of attraction $M(C_1)$ (grey) and $M(C_2)$ (white), at $x_3 = 0.0092$. The two spirals of the basins are focused in the saddle points $\mathbf{x}^*$ and $\mathbf{x}^{**}$.

When both oscillators stay in a similar attractor, the average time difference between successive peaks of the response variable $\langle \Delta t^r \rangle = \langle t^r_{k+1} - t^r_k \rangle$ is the same as that of the drive variable $\langle \Delta t^d \rangle = \langle t^d_{k+1} - t^d_k \rangle$, and both time differences are equal to the average period of the drive oscillations $\langle T^d \rangle$. The phase difference between the kth peaks of the two coupled variables can be calculated as

$$\Delta\varphi_k = \varphi^r_k(t) - \varphi^d_k(t) = 2\pi(t^r_k - t^d_k)f^d, \tag{3.7}$$

where $\varphi^r_k(t)$ and $\varphi^d_k(t)$ are the phases of the kth peak of the response and drive systems, calculated by Equation 2.63, and f^d is the dominant frequency of the power spectrum of the drive oscillator.

Without coupling, or for a very weak coupling strength, the drive signal has no effect on the response system, so that $C_2 \neq C_1$, and $\Delta\varphi$ increases linearly with time. As the coupling strength is increased, a complex scenario of synchronization occurs, which is described below.

Multistate Intermittency

A critical value of the coupling strength $\sigma_1 = 0.00505$ (referred to as *first threshold* for multistate intermittency) exists, above which the response system starts to switch repeatedly between the two coexisting chaotic states. At such small coupling, the driving signal x_2 does not change the global structure of the phase space. The time series corresponding to these switches is shown in Figure 3.5(a).

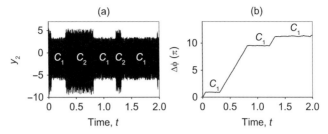

Figure 3.5 Dynamics of the response oscillator in Equation 3.1 for $\sigma = 0.0051$ (just above the first threshold for multistate intermittency). (a) Time series demonstrating random switches between the states C_2 and C_1. The state with lower amplitude variation is similar to the attractor of the drive oscillator C_1. (b) Time dependence of phase difference between response and drive systems. $\Delta\varphi$ increases linearly when the systems are in distinct states, and fluctuates around a certain average value when they are in the same state C_1.

Although no synchronization is observed, this stage can be considered as a precursor to actual synchronization, because the response system senses the drive signal, which acts as external noise and induces random escapes from the state C_2 to the state C_1 every time the trajectory closely approaches the basin boundary.

Time intervals, during which the response oscillator stays in a similar state as the drive, alternate with intervals when they stay in distinct states (Figure 3.5(b)). Similar intermittent dynamics are observed in nonidentical coupled systems.

The phase difference $\Delta\varphi$ within the windows where both trajectories belong to the same attractor C_1 drifts as a random walk with range 2π (Figure 3.6(a)), and obeys a Gaussian distribution with mean $-\pi \bmod 2\pi$ (Figure 3.6(b)). At this stage, $\Delta\varphi$ is independent of the coupling σ.

Imperfect Intermittent Phase Synchronization

As the coupling strength is increased over a *second threshold* ($\sigma_2 = 0.01$), the phase of the drive signal entrains that of the response system, while their amplitudes remain uncorrelated. This is the weakest type of synchronization, known as *imperfect intermittent phase synchronization*. The oscillators are in a phase-synchronized state only within the windows where the response and drive systems stay in the same chaotic attractor. Out of these windows, the oscillators are asynchronous.

As the coupling is further increased, the windows of phase synchronization occur increasingly frequently, while the mean width of these windows decreases. The response system is now sensitive not only to random pulses of the drive, which induce switches between the coexisting states, but also to the phase of the drive oscillations.

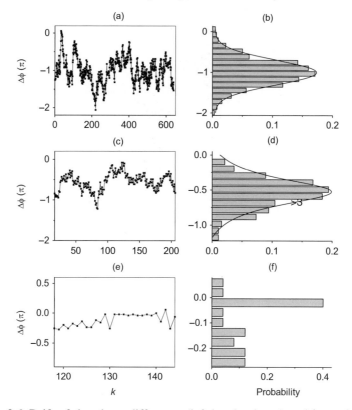

Figure 3.6 Drift of the phase difference (left-hand column) and its probability distribution (right-hand column) in windows where drive and response systems stay in a similar chaotic state. (a,b) $\sigma = 0.0051$. Random switches, no synchronization. Random walk of the phase difference inside the time window $1.33 < t < 1.95$, where both oscillators stay in state C_1. The phase drift around $\Delta\varphi = -\pi$ (over the whole phase range $\pm\pi$) is well approximated by a Gaussian distribution, indicating anticipation of the response over the driving signal. (c,d) $\sigma = 0.02$. Imperfect intermittent phase synchronization with phase drift $\Delta\varphi \in [0, -\pi]$. The phase distribution is well approximated by a Lorentzian fit. (e,f) $\sigma = 0.05$. Perfect intermittent phase synchronization with $\Delta\varphi = 0$ during some time.

Since the mean duration of the phase-synchronized windows $\langle t_L \rangle$ (in the laminar phases) increases with σ, the phase synchronization state becomes increasingly stable. At this stage, the phase difference obeys a Lorentz distribution (see Figure 3.6(d)). Note that imperfect intermittent phase synchronization is also observed in nonidentical coupled chaotic oscillators (Chen et al. 2001).

Intermittent Phase Synchronization

At larger σ, there is another critical value $\sigma_3 = 0.028$, which we will refer to as the *third threshold*. At this stage, the response system is sensitive to every peak of the

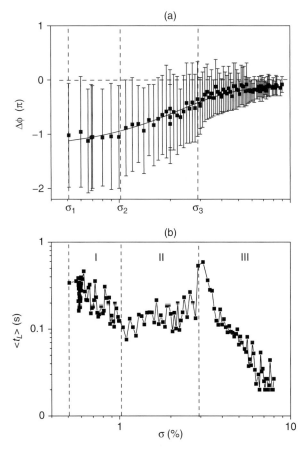

Figure 3.7 (a) Drift of phase difference (bars) and its average (dots), and (b) mean duration of laminar phase $\langle t_L \rangle$ versus the coupling strength. Three different types of intermittency occur in the regions I, II, and III. The vertical dashed lines indicate the thresholds σ_1, σ_2, and σ_3.

drive, and thus phase synchronization within the windows is almost perfect (see Figure 3.6(e)). The distribution of $\Delta\varphi$ is neither Gaussian nor Lorentzian. Rather, it is peaked around a most probable $\Delta\varphi$ (Figure 3.6(f)).

The mean duration of the laminar phase decreases with σ, and fully vanishes at $\sigma \approx 0.1$ because every peak of the drive produces a switch between the attractors. The average anticipation time also decreases with the coupling, as seen from Figure 3.7(a).

Within the windows of synchronous motion, the mean duration of the laminar phase $\langle t_L \rangle$ scales with the coupling strength σ as a power-law

$$\langle t_L \rangle \sim (\sigma - \sigma_c)^{P_c}, \tag{3.8}$$

where σ_c ($c = 1, 2, 3$) is the threshold coupling for different types of intermittency. Here, $\langle t_L \rangle$ is measured by taking very long time series and calculating the average time during which the response system stays in a state similar to that of the drive system.

Referring to Figure 3.7(b), one can identify three different regions. As σ is increased, $\langle t_L \rangle$ first decreases (region I), then it increases (region II), and then it decreases again (region III). In these regions the scaling exponents are $P_I = -0.5$, $P_{II} = 0.16$, and $P_{III} = -1$. There are also crossover zones between these regimes, where the scaling does not obey the above laws.

The first scaling exponent is a characteristic of type-I intermittency (Pomeau and Manneville 1980) and associated with the first saddle-node bifurcation at $\sigma_1 = 0.00505$, near which the probability distribution of $\Delta\varphi$ is very close to normal. In this regime, the drive can be considered as external noise playing the role of a random reinjector (Fujisaka and Yamada 1983).

The behavior changes drastically when the system passes through the second threshold σ_2. The distribution of $\Delta\varphi$ no longer retains the symmetry of a normal distribution and forms a Lorentzian shape, as shown in Figure 3.6(d). The origin of such a behavior can be found in 2π-phase jumps to phase synchronization near the saddle-node bifurcation at $\sigma_2 = 0.01$ (Lee et al. 1998). This phenomenon is quite general for any coupled chaotic phase-coherent system such as the Rössler system, or, in general, for systems in which a suitably defined phase increases steadily in time.

The third critical exponent of -1 is a signature of on-off intermittency resulting in intermittent jumps between the two coexisting states near the saddle-node bifurcation at $\sigma_3 = 0.028$. For $\sigma > \sigma_3$ the response system becomes sensitive to the shape of the drive oscillations leading to almost perfect phase synchronization inside the windows. In other words, the drive oscillations lock the oscillations of the response system, similar to a 1:1 frequency locking. The probability distributions of the laminar phase versus the laminar length for type-I and on-off intermittencies obey scaling laws with exponents $-1/2$ and $-3/2$ (Heagy et al. 1994) respectively. Note that the transition from phase-unlocked on-off intermittency to phase-locked intermittency is also observed in coupled chaotic identical monostable systems (Zhan et al. 2001).

Anticipating Synchronization

The oscillations of the response system can have a negative phase shift with respect to the drive. In other words, they can anticipate the oscillations of the drive. This means that the response system is synchronized with the future state of the drive.

Such a highly nontrivial phenomenon, known as *anticipating synchronization*, is usually observed in chaotic systems with time-delayed feedback (Voss 2000) and in

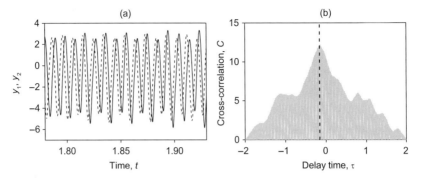

Figure 3.8 Anticipating synchronization within the intermittent phase synchronization window at $\sigma = 0.02$. (a) Time series, and (b) cross-correlation function. The solid and dashed lines correspond to the drive and response oscillations, respectively. The maximum correlation occurs for a negative time delay.

systems with frequency mismatch when the natural frequency of the response oscillator is higher than that of the drive system (see Section 2.6). However, the same phenomenon also occurs in unidirectionally coupled identical multistable systems without any delay (see Figure 3.8(a)).

The origin of anticipation in coupled systems with coexisting attractors is still an open problem. In the system of Equation 3.1, anticipating synchronization is in fact observed only for relatively small coupling ($\sigma < 0.1$) within the windows of phase synchronization, and therefore a more appropriate term to indicate this type of synchronization would be *anticipating intermittent phase synchronization*. The anticipation process is confirmed by the fact that the maximum of the cross-correlation occurs for negative delay time τ, as shown in Figure 3.8(b).

Anticipation may be understood as the fact that a small coupling acts as a small change in initial conditions of the response system, directing its phase trajectory to a future state of the drive. The anticipation time cannot be larger than the averaged period of the chaotic oscillations $\langle T^d \rangle$.

Generalized Synchronization

The last synchronization state, occurring for $\sigma \gtrsim 0.1$, is that of *generalized synchronization*. In coupled systems with coexisting attractors, generalized synchronization may take different forms, such as intermittent complete synchronization, intermittent period-doubling synchronization, period-doubling phase synchronization, and intermittent frequency locking. All these synchronization states are illustrated in Figure 3.9 with times series, phase portraits, and power spectra.

The first regime, observed for $0.1 < \sigma < 0.25$, is that of intermittent complete synchronization. It is characterized by random jumps of the coupled system from one state to another, as shown in Figure 3.9(a). The response system states

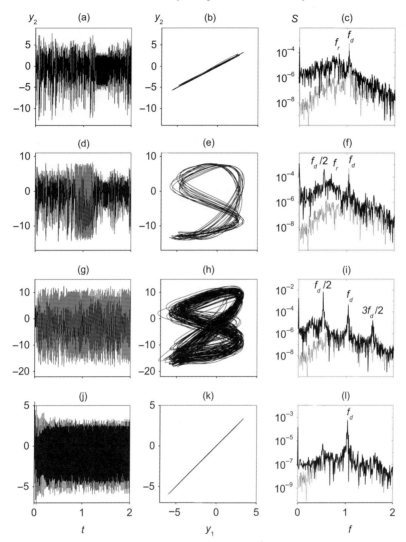

Figure 3.9 Time series (left-hand column), phase portraits inside synchronous windows (middle column), and power spectra (right-hand column) demonstrating different synchronization states for (a,b,c) $\sigma = 0.25$, intermittent complete synchronization with 1:1 phase-locking window, (d,e,f) $\sigma = 0.32$, intermittent period-doubling synchronization, (g,h,i) $\sigma = 0.62$, period-doubling phase synchronization, and (j,k,l) $\sigma = 0.7$, complete synchronization. The dashed and solid lines in the power spectra correspond to the drive and response systems with dominant frequencies f_d and f_r, respectively.

are characterized by their dominant frequency. When the response is synchronized with the drive, its dominant frequency f_d is the same as that of the driving system. When, instead, it is not synchronized, it has a dominant frequency $f_r \neq f_d$, as shown in Figure 3.9(a,b,c).

For stronger coupling, $0.25 < \sigma < 0.6$, intermittent period-doubling phase synchronization is observed. There, the drive induces a period-doubling bifurcation in the response system when it is in a state similar to that of the driving attractor. As a result, windows with period-doubled oscillations appear in the time series (Figure 3.9(d)). Within these windows, the response is phase-synchronized with the drive at the 1:2 ratio, as also seen from the phase portrait in Figure 3.9(e). The response chaotic attractor with dominant frequency f_r undergoes a methamorphosis, so that f_r moves towards half of the driving frequency as σ is increased, as shown in Figure 3.9(f).

Period-doubling phase synchronization, or 1:2 frequency locking, eventually occurs at $\sigma \approx 0.6$. At this value of the coupling, the two coexisting chaotic states converge into a single chaotic state with dominant frequency $f_d/2$. As a result, intermittency disappears and the 1:2 locking regime becomes stable (see panels g, h and i in Figure 3.9). Note that the locking of the response dominant frequency to half of the drive dominant frequency can also occur in periodically forced systems, as, for example, in a chaotic fiber laser with pump modulation (Pisarchik and Barmenkov 2005).

Intermittent frequency locking is a synchronization state when a 1:2 frequency locking regime alternates with a 1:1 frequency locking regime. This state occurs within a relatively narrow range of the coupling strength ($0.68 < \sigma < 0.7$). In this state, both f_d and $f_d/2$ exist in the power spectrum of the response system.

As the coupling is further increased, the $f_d/2$ spectral component decreases, giving preference for f_d, and finally, for $\sigma > 0.7$, the coupled system becomes completely synchronized. The dominant frequency of the response oscillator is locked by the drive at the 1:1 ratio ($f_r = f_d$) as shown in Figures 3.9(j)–3.9(l).

3.2 Systems with a Common External Force: Crowd Formation

A simple social network can be formed by three dynamical units coupled in an auxiliary configuration, as shown in Figure 2.3(d). One of the subsystem (the leader) acts as an external force, and drives two other subsystems that may represent either individuals or homogeneous groups of people, e.g., social communities, families, nations, or companies (Figure 3.10).

One can further assume that the individuals are modeled by identical multistable systems. The leader can influence the followers, and change their opinions or emotions via different communication media (TV, radio, newspapers, Internet, etc.). The individuals have in general no influence on the leader, i.e., the coupling between the leader and the individuals is unidirectional, with coupling strength K_1 that can be used as a control parameter.

Figure 3.10 Simple social network formed by a leader and two groups of individuals (e.g., opposition and sympathizers). K_1 and K_2 are the coupling strengths of the groups with the leader and the groups between themselves, respectively. The groups are described by identical multistable systems, which, if uncoupled, stay in different chaotic attractors, indicating that the individuals in each group have different opinions or emotions.

Moreover, the individuals may communicate with each other, i.e., their groups are connected bidirectionally, with coupling strength K_2. In general, the role of the leader in creating a crowd may be played not only by a person or a social group (a government, a political party), but also by a social event (e.g., a football game, a concert, or a telecast) that affects the emotions of the individuals. In the latter case, the leader's dynamics may be described by distinct parameters, or even by a distinct dynamical system.

Mathematically, a crowd can be interpreted as a state of multistable systems, where the individual systems lose their initial attractors and synchronize in a new monostable state, not necessarily identical to one of the states of the leader. Next, we show how multistable chaotic systems (individuals) synchronize under the influence of a common force (the leader), which may be either chaotic or random. We say that a leader brings people under total control (all individual trajectories follow the leader) when the coupled subsystems are completely synchronized with each other and with the drive. This is possible only if the leader and the individuals are modeled by identical equations.

When accounting for unidirectional coupling between the leader and the individuals and bidirectional coupling among the individuals, the system sketched in Figure 3.10 can be represented by the following generic differential equations

$$\dot{\mathbf{w}} = \mathbf{L}(\mathbf{w}),$$
$$\dot{\mathbf{u}} = \mathbf{I}(\mathbf{u}) + K_1\mathbf{M}(\mathbf{w} - \mathbf{u}) + K_2\mathbf{M}(\mathbf{v} - \mathbf{u}), \tag{3.9}$$
$$\dot{\mathbf{v}} = \mathbf{I}(\mathbf{v}) + K_1\mathbf{M}(\mathbf{w} - \mathbf{v}) + K_2\mathbf{M}(\mathbf{u} - \mathbf{v}),$$

where $\mathbf{w}(t)$, $\mathbf{u}(t)$, and $\mathbf{v}(t)$ are the state vectors of the drive and two response systems, $\mathbf{L}$ and $\mathbf{I}$ are their vector functions, K_1 and K_2 are the coupling strengths, and

M is the coupling matrix. Since all individuals are multistable systems, their phase trajectories may belong initially to different attractors.

We first consider the case of no communication between individuals ($K_2 = 0$). Under the influence of the drive, both response subsystems may undergo metamorphosis (Afraimovich et al. 1986) resulting in a new homeomorphic system

$$\dot{\mathbf{u}} = \mathbf{I_h}\,(\mathbf{u},\,K_1) + K_1\mathbf{Mw},$$
$$\dot{\mathbf{v}} = \mathbf{I_h}\,(\mathbf{v},\,K_1) + K_1\mathbf{Mw}, \tag{3.10}$$

where $\mathbf{I_h}\,(\mathbf{u},\,K_1) = \mathbf{I}\,(\mathbf{u}) - K_1\mathbf{Mu}$ and $\mathbf{I_h}\,(\mathbf{v},\,K_1) = \mathbf{I}\,(\mathbf{v}) - K_1\mathbf{Mv}$.

Since the second term in the definition of $\mathbf{I_h}$ is negative, it increases the system dissipation, leading to a change in the global structure of the phase space, so that, for a large enough K_1, the system becomes monostable.

Generalized synchronization occurs between the response and the drive systems, $\mathbf{u}(t) = \mathbf{H}(\mathbf{w}(t))$ and $\mathbf{v}(t) = \mathbf{H}(\mathbf{w}(t))$, when their phase variables become functionally dependent via some function $\mathbf{H}$, while $\mathbf{u}(t) = \mathbf{v}(t)$, i.e., the individuals are completely synchronized. In this case, both the originally null conditional Lyapunov exponents become negative (Guan et al. 2006). If the drive and response systems are identical ($\mathbf{L} = \mathbf{I}$), a strong enough coupling K_1 results in complete synchronization with the leader ($\mathbf{w} = \mathbf{u} = \mathbf{v}$).

A different behavior may occur when the response subsystems are coupled by common noise (e.g., an irregular, random propaganda). It is known that small noise in a multistable system can induce preference for one of a set of coexisting attractors (Kraut et al. 1999; Pisarchik and Jaimes-Reátegui 2009; Martínez-Zérega and Pisarchik 2012), whereas large noise can provoke intermittent jumps between coexisting states, (Huerta-Cuellar et al. 2008) or even stabilize a fixed point (Anishchenko et al. 2001).

3.2.1 A Numerical Example

Let n and m be two social groups and L be a leader. Suppose the dynamics of the social groups are described by piecewise linear Rössler-like equations:

$$\dot{x}_{n,m} = -\alpha x_{n,m} - \beta \left[y_{n,m} + K_1 \left(y_L - y_{n,m} \right) + K_2 \left(y_{m,n} - y_{n,m} \right) \right] - z_{n,m},$$
$$\dot{y}_{n,m} = x_{n,m} + \gamma \left[y_{n,m} + K_1 \left(y_L - y_{n,m} \right) + K_2 \left(y_{m,n} - y_{n,m} \right) \right], \tag{3.11}$$
$$\dot{z}_{n,m} = g \left(x_{n,m} \right) - z_{n,m},$$

where g is given by Equation 3.2. Here, $x_{n,m}$, $y_{n,m}$ and $z_{n,m}$ are the state variables described the dynamics of the social groups n and m, and y_L is the coupled variable associated with the leader L. When uncoupled ($K_1 = K_2 = 0$), for the choice of parameters $\alpha = 0.05$, $\beta = 0.5$, $\gamma = 0.266$, and $\mu = 15$, the systems n and m generate either Rössler chaos or homoclinic chaos, depending on the initial conditions.

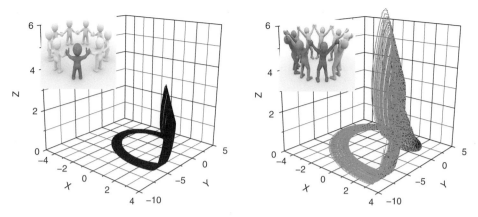

Figure 3.11 Two different opinions associated with two coexisting chaotic attractors.

Choosing the initial conditions so that the oscillators associated with the individuals stay initially in different chaotic regimes (Figure 3.11) allows us to model the coexistence of two chaotic attractors associated with different opinions.

Different types of leadership (person, event, or noise) can be represented by corresponding leader dynamics governed by the variable y_L. For simplicity, we suppose that all persons are identical. Therefore, when the leader is a person, its dynamics are modeled by the same equations as for the groups of individuals:

$$\dot{x}_L = -\alpha x_L - \beta y_L - z_L,$$
$$\dot{y}_L = x_L + \gamma_L y_L, \qquad\qquad (3.12)$$
$$\dot{z}_L = g(x_L) - z_L,$$

where $\gamma_L = \gamma$ and g has the usual form of Equation 3.2.

When the leader is instead supposed to be an event, one of the parameters is chosen to be different. For our example, we set $\gamma_L = 0.146$. Finally, the effect of noise is modeled using $y_L = D\xi(t)$, where $\xi(t)$ is zero-mean normal noise uniformly distributed within the interval $[-1, 1]$ and D is the noise intensity normalized to the maximum amplitude y_m^{max} of the uncoupled m oscillator.

Figure 3.12 illustrates the emergence of a collective state (crowd) under the influence of the drive associated with a social event, when a leader's attractor differs from individuals' attractors. When the coupling with the leader increases up to $K_1 = 0.15$ and a small coupling between the individuals is added ($K_2 = 0.04$), all individuals lose their original attractors and flow to a new chaotic state (*crowd state*), where they are completely synchronized. At the same time, the crowd is in a generalized synchronization state with the leader. When, instead, the leader is a person, a much lower K_1 is needed to create a crowd ($K_1 = 0.09$), even without

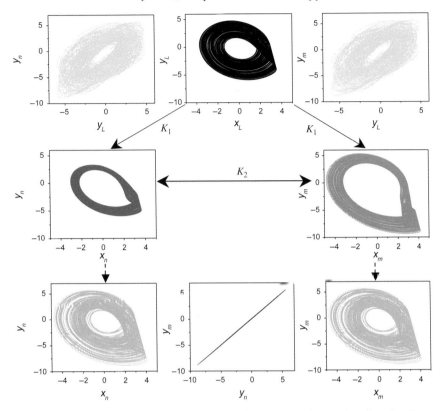

Figure 3.12 Chaotic leader. The upper middle frame shows the chaotic phase portrait of the leader. The middle frames illustrate distinct chaotic attractors of individuals. The dashed arrows show new chaotic monostable states created in the individuals by couplings $K_1 = 0.15$ and $K_2 = 0.04$. The straight line in the lower central frame means complete synchronization between the individuals. The upper left and right frames demonstrate generalized synchronization between the leader and the individuals.

communication between individuals ($K_2 = 0$). This state is referred to as a *diffuse crowd*.

The dynamics of a crowd created by noise are different. In our system we observe that a relatively strong noise ($K_1 = 0.3$) stabilizes the fixed points of the individual oscillators and all individual trajectories become equivalent in this noisy fixed point, resulting in noise-induced synchronization (Hramov et al. 2006; Guan et al. 2006). This state, illustrated in Figure 3.13, may be treated as a *quiet crowd*. Also in this case, the individuals stay in the generalized synchronization state with the leader.

Keeping $K_2 = 0$ while increasing K_1 allows one to conduct an instructive study of the phenomenology of this system, identifying several important stages in the evolution of collective behavior:

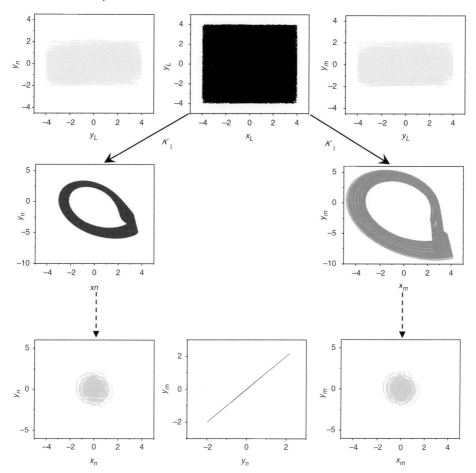

Figure 3.13 Stochastic leader. Being coupled by noise (upper central portrait), the individuals lose their initial chaotic attractors (middle portraits) and are attracted to noisy fixed points (lower left and right portraits), where they are completely synchronized (lower central portrait), while being in the generalized synchronization state with the leader (upper left and right portraits). $K_1 = 0.3$.

(i) For $0.002 < K_1 < 0.02$, random switches between coexisting states occur. This state, a precursor of synchronization, can be considered as describing changes in people's opinions, such as when they are challenged or questioned in their beliefs.

(ii) For $0.02 < K_1 < 0.15$, intermittent synchronization between individuals occurs when they are found in similar states. Continuing with the previous analogy, it is as if challenges and questions simultaneously affect many people.

(iii) For $0.15 < K_1 < 0.25$, the individuals are synchronized in phase, while their amplitudes are asynchronous. This is a precursor state of a crowd, with the individuals already having the same opinion but still acting asynchronously.

(iv) For $0.25 < K_1 < 0.5$, one reaches complete synchronization between individuals, but not necessarily with the leader. At this stage, we are dealing with crowd behavior: the individuals have the same opinion and act synchronously, but they are not yet synchronized with the leader.

(v) Finally, for $K_1 > 0.5$ complete synchronization with the leader occurs in identical systems, as when a leader has total control over the people. Increasing K_2 improves synchronization between the individuals, so that complete synchronization occurs for lower K_1, reflecting the fact that when people are together in one place and communicate with each other it is much easier for the leader to form a crowd.

Although the Rössler oscillator is not a sociological model, the main synchronization features are inherent to many multistable chaotic systems. The results obtained with Rössler oscillators coupled by an external force may explain complex sociological phenomena, such as opinion formation, crowd emergence, and leadership. In particular, they allow us to understand why pioneering theories and new products find it hard to gain widespread acceptance, or why the same idea often comes simultaneously to the minds of different people who had never previously communicated with each other.

3.3 Bidirectionally Coupled Systems

To describe the behavior of multistable systems with bidirectional coupling, we use the example of two piecewise-linear Rössler-like oscillators:

$$\dot{x}_{1,2} = -\alpha_1 \left[x_{1,2} + \beta y_{1,2} + \Gamma z_{1,2} - \sigma \psi (x_{2,1} - x_{1,2}) \right],$$
$$\dot{y}_{1,2} = -\alpha_2 \left[-\gamma x_{1,2} - (1 - \delta) y_{1,2} \right], \qquad (3.13)$$
$$\dot{z}_{1,2} = -\alpha_3 \left[-g(x_{1,2}) + z_{1,2} \right],$$

where $\psi = 20$ and $g(x)$ is given by Equation 3.2.

When uncoupled, for the choice of parameters $\alpha_1 = 500$, $\alpha_2 = 200$, $\alpha_3 = 10^4$, $\beta = 10$, $\Gamma = 20$, $\gamma = 50$, $\delta = 15.625$, and $\mu = 15$, each system exhibits the coexistence of two different chaotic attractors, like those shown in Figure 3.3. Hereafter, the attractors with larger and smaller amplitudes of the variables will be called L and S, respectively.

When the systems are coupled, four different asymptotic regimes are possible, depending on the coupling strength σ:

(i) The LL regime, in which both systems end up in an attractor indistinguishable from L.

(ii) The SS regime, in which both systems end up in an attractor indistinguishable from S.

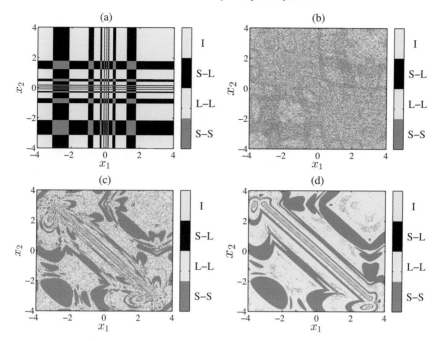

Figure 3.14 Basins of attraction of coexisting states of system Equation 3.13 for coupling strengths (a) $\sigma = 0$ (uncoupled systems), (b) $\sigma = 0.06$, (c) $\sigma = 0.14$, and (d) $\sigma = 0.20$ (reprinted with permission from Sevilla-Escoboza et al. 2015).

(iii) The SL regime, in which asymptotically one system reaches L and the other reaches S.

(iv) The I regime, in which the systems switch back and forth between L and S irregularly.

As the coupling strength σ is increased from 0 to 0.2, the four regimes appear, disappear, and mix with each other in a very complex way, depending on the initial conditions. In Figure 3.14 we show the sections of the basins of attraction of these asymptotic regimes at $y_1(0) = y_2(0) = 0$ and $z_1(0) = z_2(0) = 0$ for four representative coupling strengths, $\sigma = 0, 0.06, 0.14$, and 0.2. One can see that regime I occurs only for small coupling (Figures 3.14(b) and (c)), whereas for larger σ the oscillators synchronize and the system stays in either LL or SS. Interestingly, the SL regime is never observed for nonzero coupling.

For a better visualization of the prevalence of the different asymptotic regimes at different values of σ, Figure 3.15(a) shows the fraction of the initial conditions that eventually lead to each of the four asymptotic regimes, as a function of the coupling strength. As seen in the figure, most initial conditions lead to the intermittent behavior I at small coupling, whereas increasing σ results in the emergence of SS and LL, together with the disappearance of I.

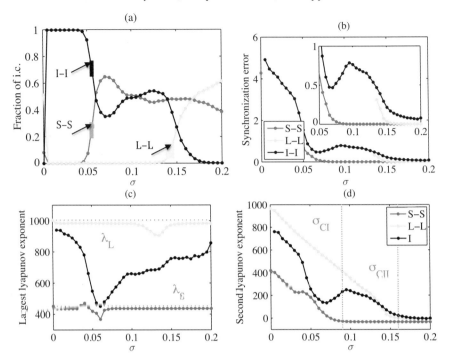

Figure 3.15 (a) Relative size of basins of each attractor, (b) synchronization error, (c) first and (d) second largest Lyapunov exponents as a function of coupling σ. See text for definitions of σ_{CI} and σ_{CII}

The synchronization error for each of the asymptotic states, shown in Figure 3.15(b), indicates that synchronization is first achieved for the SS regime around $\sigma = 0.075$, while it occurs for the LL case only for $\sigma \approx 0.165$. As for the I regime, synchronization depends nonmonotonically on the coupling. In this latter regime, the transition to synchronization passes through two stages: first, the two coupled oscillators move to the same attractor due to an initial increase of the coupling, and later on, as the coupling is further increased, they gradually synchronize, similarly to monostable coupled oscillators.

An alternative view on the emergence of synchronization is provided by the analysis of the Lyapunov spectrum. Figure 3.15(c) shows the largest Lyapunov exponent as a function of coupling for different asymptotic regimes. The dotted lines show the largest Lyapunov exponents corresponding to the attractors L and S for the solitary (uncoupled) system. In the I regime, the value of the Lyapunov exponent is clearly correlated with the presence of LL and SS windows. However, the most important information derived from the Lyapunov spectrum concerns the second largest Lyapunov exponent, which gives indirect information on the existence of the weakest forms of synchronization.

As it is seen in Figure 3.15(d), two points (highlighted as σ_{CI} and σ_{CII}) correspond to the loss of the positive second largest Lyapunov exponent in the system. The two points indicate the existence of just one chaotic mode in the system, and therefore the onset of generalized synchronization for SS and LL, respectively.

The intermittent regime I reaches generalized synchronization as well, in a regime with frequent jumps between time windows in which both oscillators are in the same state and time windows where each oscillator is in a different state.

3.4 Synchronization of Coupled Neurons

Synchronization plays a key role in signal and information processing of brain systems (Llinás 2002; Buzsáki 2006). Nerve cells (or neurons) are important elements of a neural network to provide computational functions of the brain. Fundamentally, a single neural cell consists of a body known as a soma, which contains branches of dendrites and one axon. Biological neural networks consist of a large number of individual neurons interconnected in complex circuits via axodendritic and dendrodendritic synapses. The neuron receives information through dendrites which act as input channels. Due to ionic current propagating across the neuron's membrane inside the soma, the membrane potential changes over time producing electrical signals transmitted through the axon to other neurons.

The electrical signals generated by neurons are basically classified in three distinct states: resting, spiking, and bursting. The resting state usually occurs in the absence of stimuli, i.e., when the ionic current flowing across the neuron's membranes is zero and the membrane potential is constant. When the neuron receives a stimulus of the excitatory type, it can produce an action potential (or a spike). Relatively strong stimuli can result in successive spikes referred to as tonic spiking. Finally, the bursting behavior is characterized by successive spike trains followed by a relatively long period of quiescence.

Synchronization of coupled neurons is important for information coding and signal transmission. Partial synchronization of neuronal ensembles underlies the formation of population rhythms of the brain (alpha, theta, and gamma oscillations) (Buzsáki and Chrobak 1995; Csicsvari et al. 2000; Rex et al. 2009). At the same time, synchronous activity of large ensembles of brain neurons can be associated with pathologic processes, most notably epileptiform activity (Fisher et al. 2005).

The interest in mathematical modeling of neuron synchronization has significantly increased after real neurobiological experiments with electrically coupled neurons performed by Elson et al. (1998, 2002), in which various synchronous states have been identified. To simulate cooperative neuron dynamics, numerous models based on either iterative maps or differential equations in various coupling configurations have been developed (Chialvo 1995; Kinouchi and Tragtenberg 1996; Kuva et al. 2001; Laing and Longtin 2002; Rulkov et al. 2004; Shilnikov and Rulkov 2003; Copelli et al. 2004; Izhikevich and Hoppensteadt 2004; Sun et al. 2009; Cao and Sanjuan 2009; Elson et al. 1998; Hodgkin and Huxley 1952; Hindmarsh and Rose 1984; Ivanchenko et al. 2008; Rabinovich and Abarbanel 1999; Lang et al. 2010; Matias et al. 2011).

Synchronization of neurons were studied as a function of intrinsic and external parameters. Depending on the coupling strength and synaptic delay time, coupled neurons can give rise to spike sequences that are matching in their timings, or bursts either with lag or anticipation (Rulkov 2002; Ivanchenko et al. 2004; Lang et al. 2010; Franoviĉ and Miljkoviĉ 2010; Mayol et al. 2012). Furthermore, under specific conditions, intermittency between different synchronous states can arise (Elson et al. 2002). For example, Matias et al. (2013) demonstrated a smooth transition from lag to anticipating synchronization of coupled Hodgkin–Huxley neurons (Hodgkin and Huxley 1952) when the inhibitory synaptic conductance was increased.

In this we will summarize some features of synchronization of coupled neurons, making reference to popular neural models. In particular, we concentrate on phenomena such as phase locking and frequency locking, lag and anticipating synchronization with memory, and the emergence of phase synchronization of neurons with arbitrary phase shift.

3.4.1 Synchronization of Neurons with Memory

One important neuron function is information transmission through their network. This process is characterized by a certain delay time, due to a finite velocity of the action potential propagating along the neuron axon, and time lapses in dendritic and synaptic processes (Kandel et al. 2000). The delay in synaptic connections (Katz and Miledi 1965) is required for a neurotransmitter to be released from a presynaptic membrane, diffuse across the synaptic cleft, and bind to a receptor site on the postsynaptic membrane.

A common structural feature of neural networks, long predicted to be responsible for short-term memory (Amit 1995), is the presence of feedback loops. A schematic model of a memory cell is shown in Figure 3.16, where a feedback loop is formed in a single neuron. This system has two delays, one in the synaptic connection and another in the signal propagation along the axon. We are interested in understanding how these time delays affect synchronization of coupled neurons.

To answer this question, let us refer to one of the simplest neuron models, the Rulkov map (de Vries 2001; Rulkov 2001, 2002). The master (x, y) and slave (u, v) neurons are defined as follows:

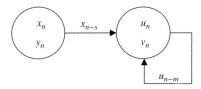

Figure 3.16 Two coupled neurons with coupling delay s and memory m.

$$
\begin{aligned}
x_{n+1} &= f\,(x_n, y_n)\,, \\
y_{n+1} &= y_n - \mu\,(x_n + 1) + \mu\sigma, \\
u_{n+1} &= f\,(u_n, v_n + \beta_n)\,, \\
v_{n+1} &= v_n - \mu\,(u_n + 1) + \mu\sigma + \mu\sigma_n,
\end{aligned}
\tag{3.14}
$$

where x_n (u_n) and y_n (v_n) are the fast and slow variables of the master (slave) neurons, β_n and σ_n relate to external stimuli, μ and σ are intrinsic parameters, and f is defined as

$$
f\,(x_n, y_n) =
\begin{cases}
\dfrac{\alpha}{1-x_n} + y_n & \text{for } x_n \leqslant 0 \\[2mm]
\alpha + y_n & \text{for } 0 < x_n < \alpha + y_n \text{ and } x_{n-1} \leqslant 0 \\[1mm]
-1 & \text{for } x_n \geqslant \alpha + y_n \text{ or } x_{n-1} > 0,
\end{cases}
\tag{3.15}
$$

where α is an intrinsic parameter.

Although this model is not explicitly inspired by physiological processes in the membrane, it features a substantially complex behavior and can reproduce specific neuron dynamics (silence, periodic spiking, and chaotic bursting), thus replicating to a great extent most of the experimentally observed regimes (Elson et al. 1998, 2002; Rulkov 2002), including spike adaptation (Rulkov et al. 2004), routes from silence to bursting mediated by subthreshold oscillations (Shilnikov and Rulkov 2004), emergent bursting (de Vries 2001), phase and antiphase synchronization with chaos regularization (Rulkov 2002, 2001), and complete and burst synchronization (Shilnikov et al. 2008; Belykh et al. 2005; Wang et al. 2008).

When the physiological response of the postsynaptic neuron to a signal is assumed to be immediate, the coupling between the cells can be defined as

$$
\beta_n = \sigma_n = \eta(x_{n-s} - u_{n-m}),
\tag{3.16}
$$

where s and m are the synaptic and memory delay times (in units of number of iterations of the map), and η is the coupling strength.

Below, we briefly describe the phenomenology of this model, using $\sigma = -0.025$, $\alpha = 5.3$ and $\mu = 0.001$ as parameters.

Frequency Entrainment

When uncoupled, both neurons are in a regime of tonic spikes. Synchronization is affected by changes in the relationship between the two delay times, and in the coupling strength η. When η is too weak, the neurons fire asynchronously. For strong enough coupling, the spikes of the slave neuron are entrained by the master neuron at a ratio p:q, where p and q are the number of spikes of the slave and master neuron, respectively. Figure 3.17 illustrates typical frequency-locked states with ratios 1:1, 2:1, and 3:1.

For intermediate values of the coupling, for which the spike frequency is not entrained, random switches between states with different frequency occur. The switches between the 1:1 and 2:1 entrained states are shown in Figure 3.18.

As the coupling strength is increased, the number of spikes in a burst of the postsynaptic neuron increases, saturating to a ratio of 13:1 at $\eta = 0.6$ (Figure 3.19(a)), and for larger coupling the slave neuron behaves chaotically, as depicted in Figure 3.19(b).

Phase Locking

As soon as the frequency is entrained, the phase of the slave neuron drifts over time, so that the difference $\Delta\varphi = \varphi_{post} - \varphi_{pre}$ between instantaneous phases of the postsynaptic and presynaptic neuron spikes fluctuates around a mean value $\vartheta = m - s$, indicating imperfect phase locking. For certain values of η the phase locking becomes perfect. Figure 3.20 shows how phase synchronization depends on the coupling strength. Perfect phase locking occurs only within narrow ranges of the coupling η.

Anticipating Synchronization

The relation between memory m and synaptic delay s is crucial for neuron synchronization. The type of synchronization is determined by the sign of ϑ. When $\vartheta < 0$ the neurons synchronize with lag, otherwise they synchronize either with anticipation (for $\vartheta > 0$) or with zero lag (for $\vartheta = 0$). Both lag and anticipated synchronization can be quantitatively characterized by the similarity function $S^2(\tau)$ (Equation 2.81), computed between the fast variables x_n and u_n. If $S^2(\tau) = 0$ for some value of τ, the neurons are synchronized, either with delay or anticipation.

A typical regime of anticipating synchronization is shown in Figure 3.21. One can see that the spikes of the slave neuron anticipate the spikes of the master neuron with anticipation time $\vartheta = m - s = 10$.

3.4.2 Synchronization of Neurons with Arbitrary Phase Shift

On the local dynamics level, the synchronous activation of a pair of neural oscillators with a small phase shift between the interacting neurons is responsible for

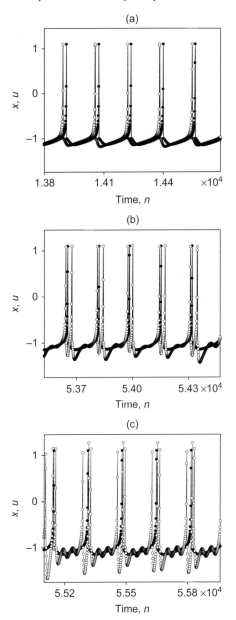

Figure 3.17 Time series of fast variables of two coupled Rulkov maps, demonstrating frequency entrainment to (a) 1:1 ratio for $m = 16$, $s = 4$, and $\eta = 0.04$, (b) 2:1 ratio for $m = 1$, $s = 0$, and $\eta = 0.009$, and (c) 3:1 ratio for $m = 1$, $s = 0$, and $\eta = 0.02101$. The closed and open dots correspond to the master and slave neuron spikes, respectively.

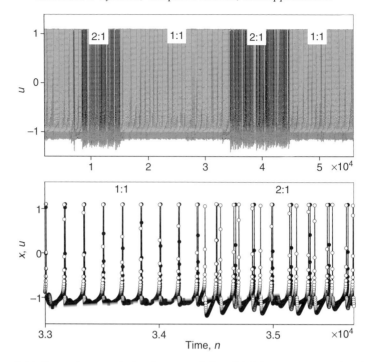

Figure 3.18 Intermittent switches between 1:1 and 2:1 frequency-entrained states for $\eta = 0.00401$, $m = 1$, and $s = 0$. The lower frame is an enlarged part of the upper frame.

long-term changes in the efficiency of signal transmission (Markram et al. 1997; Bi and Poo 1998). This phenomenon is thought to underlie learning and cognitive functions in the brain, and its properties are phase selective, implying a dependence of the parameters of the neural network on the relative phase of spikes (Kazantsev and Tyukin 2012).

In classical works (Ermentrout and Kopell 1990; Hansel et al. 1993), the analysis of synchronization is reduced to the study of simple phase equations. The calculation of the phase response curve for various external actions made it possible to reveal stable phase-locking regimes, and to establish their relationships with the type of bifurcation in the oscillator model (Hansel et al. 1995; Ermentrout et al. 2012) and with the presence of resonance (Izhikevich 2007). In most studies, the coupling strength and depolarization level are used as the main control parameters.

We focus here on stable phase-locking regimes with different phase shifts in a pair of synaptically coupled spiking neural oscillators. A representative feature of the model is the dependence of the steady-state phase on the applied current that determines the neuron depolarization level, which can be controlled in neural

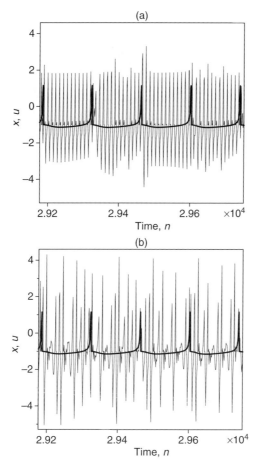

Figure 3.19 Time series for coupling (a) $\eta = 0.6$ and (b) $\eta = 0.8$. The thick and thin curves show respectively the spikes of the master and slave neurons. The increasing number of spikes in bursts results in irregularities in the spiking sequence of the slave neuron.

cells by means of extracellular signals. The relative phase is independent of the coupling strength between the neural oscillators, and anticipating synchronization is possible under specific conditions (Simonov et al. 2013).

Consider two spiking neurons with arbitrary phase shift, unidirectionally coupled via a chemical synapse. The neuron dynamics are modeled by the Hodgkin–Huxley equations (Hodgkin and Huxley 1952) for the membrane potentials V_1 and V_2:

$$
\begin{aligned}
C\dot{V}_1 &= I_{ion,1} + I_{app,1}, \\
C\dot{V}_2 &= I_{ion,2} + I_{app,2} + I_{syn},
\end{aligned}
\tag{3.17}
$$

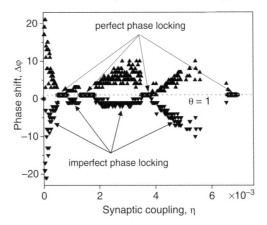

Figure 3.20 Phase difference (in units of number of iterations) versus coupling strength for $\vartheta = 1$. While the slave neuron frequency is entrained to 1:1, the phase difference drifts in time (imperfect phase locking) around its mean value $\vartheta = 1$ (shown by the dotted line). Only for certain η, the phase is locked with anticipation time ϑ.

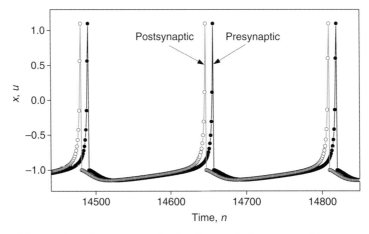

Figure 3.21 Anticipating synchronization in coupled neurons with memory. $\nu = 0.055$, $m = 11$, $s = 1$, $\vartheta = 10$. The slave neuron fires before the master neuron.

where subscripts 1 and 2 mark respectively the presynaptic (master) and postsynaptic (slave) neurons and C is the specific membrane capacitance. I_{ion} is a sum of ionic currents which includes sodium (Na), potassium (K), and leak (L) currents, given as

$$I_{ion,1,2} = -g_{Na}m^3 h\left(V_{1,2} - E_{Na}\right) - g_K n^4\left(V_{1,2} - E_K\right) - g_{Na}\left(V_{1,2} - E_L\right), \quad (3.18)$$

where the parameters $x = m, n, h$ are solutions of the equation

$$\dot{x} = \alpha_x(1 - x) - \beta_x x, \quad (3.19)$$

in which α and β depend on the membrane potential. The values for the gating parameters g_{Na}, g_K, and g_L, and for the potentials E_{Na}, E_K and E_L can be found in the literature (see, for example, Izhikevich 2007). The applied currents $I_{app,1,2}$ are constant and determine the depolarization level of neurons, as well as the dynamical regime (excitable, oscillatory, or bistable) and the spike generation frequency in oscillatory regimes.

The action of the master neuron on the slave neuron is described by the synaptic current I_{syn} in the equation for the membrane potential of the slave neuron. The inhibitory synaptic current I_{syn} is assumed to have the form

$$I_{syn} = \frac{g_{syn}\left(V_2 - V_{syn}\right)}{1 + e^{\frac{\vartheta_{syn}-V_1}{k_{syn}}}}. \tag{3.20}$$

The reversal synaptic potential for the excitatory connection is $V_{syn} = 0$. This means that the postsynaptic neuron in the resting state acquires a negative synaptic current (usually $V_2 \approx -70$ mV). This determines the rate of increase in the membrane potential and, therefore, the spike generation probability. The threshold synaptic function ϑ_{syn}, specifying its shift, and the steepness k_{syn} are chosen to be $\vartheta_{syn} = 0$ and $k_{syn} = 0.2$ mV, to ensure the short-term response of the postsynaptic neuron only to the top of the presynaptic pulse. The action of the subthreshold fluctuations of the membrane potential of the master neuron is cut off.

The *relative spiking phase* in the case of coupled neural oscillators is defined as the difference between the timings of spike generation by two oscillators, t_2^{sp} and t_1^{sp}, divided by the oscillation period T (Kazantsev et al. 2005):

$$\varphi(n) = \frac{t_2^{sp}(n) - t_1^{sp}(n)}{T}, \tag{3.21}$$

where n is the index of the postsynaptic spike.

Figure 3.22 shows the membrane potentials (left panels) of the model and the corresponding phase dynamics (middle panels) computed by Equation 3.21. One can observe antiphase (a), in-phase (b), and anticipating (c) synchronization depending on the parameters. The phase difference φ_n between presynaptic and postsynaptic spikes is determined by the initial conditions for the slave oscillator at the limit cycle. The phase of the presynaptic neuron is chosen to be zero at the point where the potential has a maximum.

3.4.3 Phase Map

Phase dynamics of coupled neurons can be studied using a one-dimensional spiking phase map

$$\varphi(n + 1) = F(\varphi(n)), \tag{3.22}$$

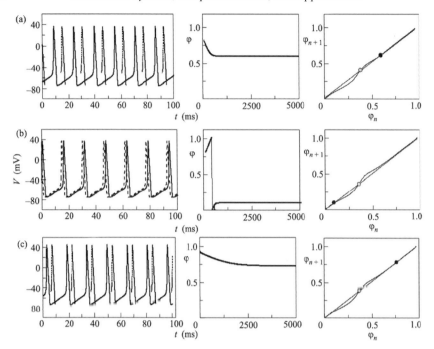

Figure 3.22 Examples of synchronization regimes in the model of a pair of synaptically coupled spiking neural oscillators Equation 3.17: (a) antiphase regime, (b) in-phase regime with delay, and (c) anticipating regime. The left-hand panels show the membrane potentials after transients for the presynaptic (dashed lines) and postsynaptic (solid lines) neurons. The middle panels show the time dependence of the relative phase. The right-hand panels represent the phase maps, where the closed and open circles mark the stable and unstable fixed points, respectively (reprinted with permission from Simonov et al. 2013).

where φ is the phase difference computed via Equation 3.21, and F is a function that describes the phase shift for the next incoming spike.

Examples of phase maps corresponding to in-phase and antiphase synchronization are presented in the right-hand panels in Figure 3.22. The intersections of the phase map curve with the diagonal line yield the fixed points. One can see that in all cases there are one stable and one unstable fixed point. Each stable fixed point corresponds to a steady relative phase, and all initial conditions (excluding the unstable fixed point) converge to the stable fixed point defining the phase-locked oscillations.

Alternatively, for weak enough couplings, one can construct the phase response curve, by plotting the change in relative spiking phase $\Delta\varphi = \varphi(n+1) - \varphi(n)$ as a function of $\varphi(n)$ (Goel and Ermentrout 2002; Guevara et al. 1986; Kuramoto 1984; Hansel et al. 1993; Netoff et al. 2005). The fixed points are the values of

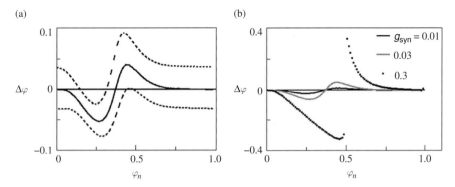

Figure 3.23 Phase response $\Delta\varphi$ for the Hodgkin–Huxley model. Panel (a) shows $\Delta\varphi$ at fixed $I_{app,1} = -8\ \mu A/cm^2$ and $g_{syn} = 0.025\ mS/cm^2$ for $I_{app,2} = -7.3\ \mu A/cm^2$ (upper line), $8\ \mu A/cm^2$ (middle line), and $8.7\ \mu A/cm^2$ (lower line). Panel (b) shows $\Delta\varphi$ at $I_{app,1} = I_{app,2} = 8\ \mu A/cm^2$ for $g_{syn} = 0.01\ mS/cm^2$ (dark line), $0.03\ mS/cm^2$ (light line), and $0.3\ mS/cm^2$ (dotted line) (reprinted with permission from Simonov et al. 2013).

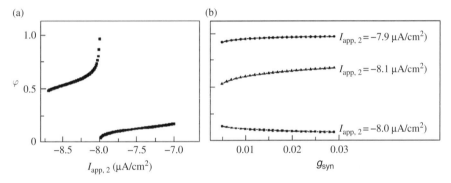

Figure 3.24 Relative spiking phase φ of the stable fixed point. Panel (a) shows φ as a function of $I_{app,2}$ for a coupling $g_{syn} = 0.025\ mS/cm^2$. Panel (b) shows φ as a function of g_{syn} for different values of $I_{app,2}$. In both cases it is $I_{app,1} = -8\ \mu A/cm^2$ (reprinted with permission from Simonov et al. 2013).

$\varphi(n)$ for which $\Delta\varphi = 0$. The derivative of $\Delta\varphi$ is negative at stable fixed points and positive at unstable ones.

In the Hodgkin–Huxley model of Equation 3.17, $\Delta\varphi$ is affected both by the coupling strength g_{syn} and by the detuning induced by the depolarization level of the postsynaptic neuron $I_{app,2}$, as shown in Figure 3.23. Note that high couplings can cause discontinuities in $\Delta\varphi$, as evident from Figure 3.23(b). To better appreciate the effect of these parameters, in Figure 3.24 we plot the stable fixed point as a function of $I_{app,2}$ and g_{syn}. It is clear that the key parameter that determines the stable phase value is the depolarization level, rather than the coupling strength. Thus, if the coupling is reasonably small, an arbitrary stable phase can be selected by choosing an appropriate detuning.

The fact that two oscillating neurons can establish at certain conditions either anticipating or lag synchronization is believed to be of importance in large neural networks, where numerous synchronized pairs can control the switching between signal propagation paths via synaptic rearrangements initiated by adjustments of their depolarization levels (Markram et al. 1997).

3.5 Chaos Synchronization for Secure Communication

As a relevant application of what discussed so far, we have chosen to briefly describe the use of chaos synchronization for secure communication purposes.

We currently live in the era of information, and societies, commerce, governments and private institutions require secure and fast communication. With the growing demand of electronic commerce, society in general needs to keep its information safe while traveling through the communication network at high speeds.

In this aspect, chaos has been widely considered as a viable option in cryptography for secure communication, because many of its fundamental characteristics, such as a broadband spectrum, ergodicity, and high sensitivity to initial conditions, are directly connected with two basic properties of good ciphers: confusion and diffusion.

Diffusion is associated with dependence of the output on the input bits, and as such it is directly correlated to the property of redundancy, which allows the statistics of plaintext (original information) to be dissipated in the statistics of ciphertext (encrypted information); confusion refers to making the relationship between the key and the ciphertext as complex and involved as possible.

The greatest advantage of a chaotic system over a noisy one is that the chaotic system is deterministic, so that the exact knowledge of initial conditions and system parameters enables one to recover the message. This property of chaos significantly facilitates the decryption process.

The idea of chaotic cryptography can be traced back to Claude Shannon (1949). Although he did not explicitly use the word "chaos," he did mention that well-mixing transformations in a good secrecy system can be constructed on the base of the stretch-and-fold mechanism, which is really a chaotic motion.

In the first scientific paper on chaotic cryptography, Matthews (1989) came up with the idea of a stream cipher based on a one-dimensional chaotic map. One year later, Pecora and Carroll (1990) published a pioneering work on synchronization of chaotic systems. Afterwards, chaotic cryptography forked over two distinct paths with almost no interaction between them: digital chaotic ciphers (Habutsu et al. 1991; Baptista 1998; Fridrich 1998) and chaos synchronization (Ashwin 2003; Argyris et al. 2005; Pisarchik 2008; Pisarchik and Ruiz-Oliveras 2010). The principal difference between these two approaches is that the former requires a predetermined secret key, while in the latter the key is the system itself.

The main advantage of chaotic synchronization schemes is their easy analog implementation. Traditionally, encryption is based on discrete number theory, so data have to be digitized before any encryption process can take place. When encrypting a continuous voice or a video the old-fashioned way, digitalization can pose a heavy computational burden. Using chaotic communication makes it possible to encrypt a message without the need to digitize it first. Furthermore, chaotic encryption can be implemented by means of fast analog components (electronic and/or optical).

A very important feature of any encryption scheme is its security. The traditional approach has proven to be reliable, while the security of chaotic encryption still poses some problems. The incorporation of chaotic dynamics in cryptology is a relatively new approach that only started a decade ago. Nevertheless, different cryptanalytic techniques have already been developed to estimate the security of the proposed chaotic ciphers and chaotic synchronization schemes.

Also, it is worth highlighting that no cryptosystem, with the exception maybe of quantum systems (Ekert 1991), is guaranteed to remain forever secure, as better cryptanalysis methods are always popping up. At present, quantum cryptography is still unacceptable for modern secure communication, because of two serious drawbacks: first, it is too slow and second, it can only be used over point-to-point connections and not through networks where data have to be routed. Thus, many scientists find chaotic synchronization to be a very promising paradigm for secure communication.

In the communication protocol, messages are embedded within a chaotic carrier in a transmitter and recovered after transmission in a receiver upon synchronization with the former. Special attention has been paid to optical communication using semiconductor lasers due to its direct comparability with existing optical fiber technology (Shore et al. 2008), especially after successive experiments with the Athens fiber networks (Argyris et al. 2005). In these lasers, high-dimensional chaotic signals with a large information entropy are generated by means of delayed feedback (Mirasso et al. 1996; Ruiz-Oliveras and Pisarchik 2006). The system performance largely depends on the quality of chaos synchronization, which means that it is particularly important to minimize the synchronization error.

The majority of existing communication schemes are based on complete synchronization and use only a single channel for both the laser coupling and the signal transmission. A drawback of such schemes is that complete synchronization requires the identification of the transmitter and the receiver, which is very difficult to achieve in practice.

A different secure communication system uses the concept of generalized synchronization and its combination with complete synchronization (Murali and Lakshmanan 1998). However, because of the use of a single communication channel, the signal itself creates a synchronization error that reduces the communication quality.

A further approach was introduced by Terry and VanWiggeren (2001). Their proposed setup consisted of two identical pairs of chaotic systems (master and slave), one in the transmitter and one in the receiver. A binary message is encrypted in the coupling strength between master and slave in the transmitter and recovered by analyzing the error dynamics in the receiver. In fact, this method is a modification of a chaotic shift keying with the important innovation of using two channels, one to provide complete synchronization between the transmitter and the receiver master systems, and the other to compare the two slave trajectories.

Many communication systems based on chaos synchronization have an important drawback: a binary bit message inherently produces a synchronization error, hence limiting the communication rate with the synchronization time because the transmitter and the receiver are not continuously synchronized.

Another two-channel communication scheme (García-López et al. 2005, 2008; Hernandez et al. 2008) mandates a coupling between transmitter and receiver via one of the system variables to synchronize them, while a signal is transmitted via another variable. The main advantage of this scheme is that a message added to one of the variables does not enter the receiver, thus preventing a disturbance that would inevitably give a synchronization error. Since all variables are functionally dependent in the system equations, generalized synchronization between the trajectory projections in phase space takes place; if the transmitter and the receiver are identical systems coupled by only one variable and present complete synchronization in one projection, the same will hold for any other projection. Thus, there is a combination of generalized synchronization between the system variables and complete synchronization between one of the variables of the transmitter and of the corresponding one of the receiver.

3.5.1 Communication Using Chaotic Semiconductor Lasers

A general communication system contains a transmitter and a receiver. The transmitter is in charge of modulating a signal into a series of bits, which are then sent to the receiver. The principle of applying of chaotic synchronization to communication lies in sending information over a chaotic carrier that allows signal encryption in a wide frequency range. In this case, both the transmitter and the receiver must

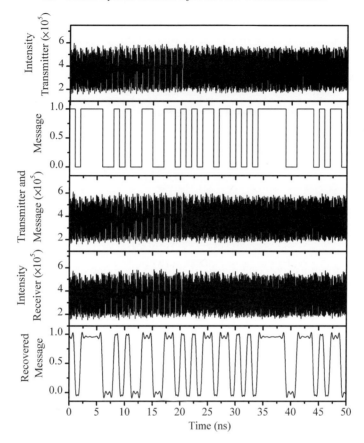

Figure 3.25 Encoding and decoding messages using chaotic carrier. The message is added to the chaos waveform in the transmitter and recovered from chaos in the receiver as the synchronization error (reprinted with permission from Pisarchik and Ruiz-Oliveras 2015).

generate chaotic waveforms. A message is added to the chaotic output of the transmitter to be recovered after the filtering in the receiver, which must be synchronized with the transmitter. Figure 3.25 illustrates how this method works.

Among many encoding and decoding schemes, we focus here on the three most popular ones, which are commonly achieved with semiconductor lasers: chaotic modulation, chaotic masking, and shift keying.

Chaotic Modulation

Chaotic modulation resembles classic amplitude modulation (AM). The message is added by modulating the emitter's chaotic carrier according to the following expression

$$M(t) = [1 - \varepsilon m(t)]P_t(t), \tag{3.23}$$

where ε is the amplitude of the encoded message, $m(t)$ is the message itself, and $P_t(t)$ is the transmitter laser intensity. In this scheme, the message $m(t)$ and the intensity $P_t(t)$ have the same phase. Since the transmitter laser is synchronized with the receiver laser, the message is recovered as follows

$$m_r(t) = \frac{1}{\varepsilon}\left[1 - \frac{M(t)}{P_r(t)}\right], \tag{3.24}$$

where $P_r(t)$ is the receiver laser intensity and $m_r(t)$ is the recovered message.

Chaotic Masking

For chaotic masking, the message is just added to the intensity of the transmitter laser as

$$M(t) = P_t + \varepsilon m(t). \tag{3.25}$$

To recover the message, one needs to subtract the intensity of the receiver laser from the incoming signal from the transmitter as follows

$$m_r(t) = \frac{M(t) - P_r(t)}{\varepsilon}. \tag{3.26}$$

Chaotic Shift Keying

In chaotic shift keying, the signal is added to the transmitter itself, but not to its output signal. This can be done by adding the message to the pump current of the transmitter laser as

$$I(t) = I + \varepsilon m(t), \tag{3.27}$$

where I is the constant pump current. To recover the message, one just needs to subtract the intensity of the transmitter laser from that of the receiver laser as follows

$$m_r(t) = P_t(t) - P_r(t). \tag{3.28}$$

Communication with chaos can be achieved by means of one-channel or two-channel communication schemes, each having advantages and drawbacks. In the following, we discuss the details of these two schemes.

3.5.2 One-Channel Communication Scheme

A one-channel communication scheme consists of a transmitting master laser (ML) and a receiving slave laser (SL), both operating in a chaotic regime. The message is encrypted into the chaotic output of the ML. To recover the message, the SL needs to be synchronized with the ML. Figure 3.26(a) illustrates this scheme for both chaotic modulation and chaotic masking, while Figure 3.26(b) shows how it works with shift keying.

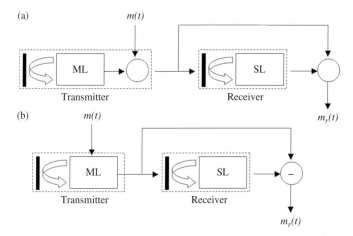

Figure 3.26 Schemes for (a) chaotic modulation and chaotic masking, and (b) shift keying (reprinted with permission from Pisarchik and Ruiz-Oliveras 2015).

Communication Quality

There are several methods to analyze the quality of a recovered signal. A common one is the eye diagram, which consists in splitting up the message into a series of fixed intervals, which are then shifted and overlapped, as shown in Figure 3.27. This is the easy way to concentrate all message bits in a small time interval.

Another measure for quantitative estimation of communication quality is the Q-factor, given by (Kanakidis et al. 2003)

$$Q = \frac{S_1 - S_0}{\sigma_1 + \sigma_0},\tag{3.29}$$

where S_1 and S_0 are the average optical intensities of bits "1"and "0," and σ_1 and σ_0 are the corresponding standard deviations.

In Figure 3.28 we show the Q-factor of a message transmitted at the rate of 1 Gb/s as a function of the coupling between the transmitter and the receiver using chaotic modulation. The squares and the circles correspond to message amplitudes yielding 2 percent and 4 percent modulation of the chaotic carrier, respectively. For good transmission, the eye diagram of the recovered message should be clean. This typically requires $Q \geqslant 10$, as illustrated in Figure 3.29.

One of the reasons why a message may not be recovered with good quality is the fact that the mean synchronization error between the transmitter and the receiver

$$\langle e \rangle = \left\langle \sqrt{(P_m(t) - P_s(t))^2} \right\rangle \tag{3.30}$$

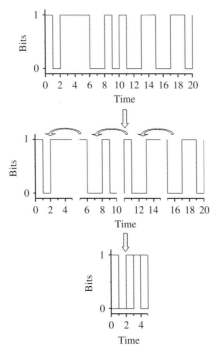

Figure 3.27 Forming eye diagram from a message by splitting it into a series of 5 bits (reprinted with permission from Pisarchik and Ruiz-Oliveras 2015).

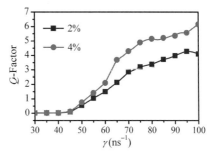

Figure 3.28 Q-factor as a function of coupling γ using chaotic modulation (reprinted with permission from Pisarchik and Ruiz-Oliveras 2015).

increases when the message is added, because the message itself acts as an external perturbation to the system. Here, P_m and P_s are the ML and SL intensities, respectively.

Figure 3.30 shows how the mean synchronization error varies with coupling strength γ for different encryption schemes, when the message is added to the transmitter. The mean synchronization error is normalized to 1 (indicating no synchronization). The squares in Figure 3.30 correspond to the mean synchronization

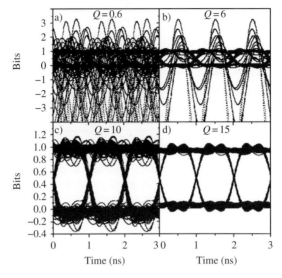

Figure 3.29 Eye diagrams and corresponding Q-factors for chaotic modulation. The diagrams are clean when $Q \geqslant 10$ (reprinted with permission from Pisarchik and Ruiz-Oliveras 2015).

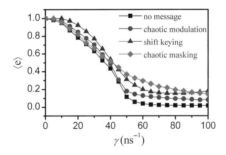

Figure 3.30 Mean synchronization error between transmitter and receiver lasers as a function of coupling, for different encryption methods (reprinted with permission from Pisarchik and Ruiz-Oliveras 2015).

error when no message is added. When $\gamma \approx 60 \text{ ns}^{-1}$, $\langle e \rangle$ decreases to a very small value, indicating complete synchronization. When the message is added, the mean synchronization error increases for all encryption methods. However, the increment for chaotic modulation is not as strong as for chaotic shift keying and chaotic masking.

3.5.3 *Two-Channel Communication Scheme*

To avoid the increase in synchronization error when the message is added, a two-channel communication system can be used (Pisarchik and Ruiz-Oliveras 2010).

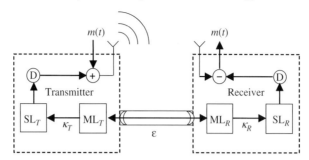

Figure 3.31 Two-channel optical chaotic communication scheme using chaos masking encryption. ML_T, SL_T and ML_R, SL_R are the master and slave lasers in the transmitter and in the receiver, respectively, κ_T and κ_R are the coupling strengths between the lasers in the transmitter and in the receiver, ε is the coupling strength between the master lasers, and D are photodetectors (reprinted with permission from Pisarchik and Ruiz-Oliveras 2015).

This consists in a transmitter and a receiver, each containing a master–slave pair. Figure 3.31 illustrates this scheme, for both chaotic masking and modulation. The master laser in the transmitter (ML_T) has the same parameters as the master laser in the receiver (ML_R) and they are coupled unidirectionally or bidirectionally via an optical fiber that acts as synchronization channel.

Since these lasers are identical, complete synchronization can be achieved for a strong enough coupling. The slave laser in the transmitter (SL_T) is identical to the slave laser in the receiver (SL_R), i.e., they have the same parameters. However, ML_T and SL_T are different, so that only generalized synchronization is possible between them. The same situation occurs for the receiver, where ML_R and SL_R are different, making only generalized synchronization possible.

The communication system shown in Figure 3.31 is based on the combination of generalized synchronization between the two pairs of master and slave lasers, and complete synchronization between ML_T and ML_R, and can be described by the following general equations:

$$
\begin{aligned}
\dot{\mathbf{X}}_T &= \mathbf{f}\left(\mathbf{X}_T, \boldsymbol{\psi}\left(\mathbf{X}_R, \varepsilon\right)\right), \\
\dot{\mathbf{Y}}_T &= \mathbf{g}\left(\mathbf{Y}_T, \mathbf{h}\left(\mathbf{X}_T, \kappa_T\right)\right), \\
\dot{\mathbf{X}}_R &= \mathbf{f}\left(\mathbf{X}_R, \boldsymbol{\psi}\left(\mathbf{X}_T, \varepsilon\right)\right), \\
\dot{\mathbf{Y}}_R &= \mathbf{g}\left(\mathbf{Y}_R, \mathbf{h}\left(\mathbf{X}_R, \kappa_R\right)\right).
\end{aligned}
\tag{3.31}
$$

Here, $\boldsymbol{\psi}$ is the coupling function between ML_T and ML_R, and it appears in both equations only if the coupling is bidirectional, κ_T and κ_R are the coupling coefficients between master and slave lasers, and ε is the coupling strength between ML_T and ML_R, which must be high enough to provide complete synchronization.

A message $m(t)$ is added to the intensity of the SL_T and the sum signal $y_T + m(t)$ is transmitted via the information channel, which can be build using any communication medium, such as a satellite antenna, an electric cable, or another optical fiber. In each moment, the slave laser is synchronized with the master laser by a certain function representing generalized synchronization, while the two slave lasers (transmitter with receiver) are completely synchronized.

When the master–slave pairs in the transmitter and in the receiver are in generalized synchronization, and ML_T and ML_R are completely synchronized, the message can be easily extracted by comparing y_R with $y_T + m(t)$ in the receiver, if the master and the slave lasers are identical and $\kappa_T = \kappa_R$, $x_T = x_R$, and $y_T = y_R$. However, real semiconductor lasers always have some mismatch in their parameters and thus cannot fulfill the identity condition. Nevertheless, even in the case of small parameter mismatch between lasers, the communication quality can be good enough when a bidirectional coupling scheme between masters is used (Pisarchik and Ruiz-Oliveras 2010).

Cross-Correlation and Synchronization Error

The quality of synchronization can be estimated by calculating the normalized cross-correlation C between the output powers of two coupled lasers i and j

$$C(t) = \frac{\left\langle P_i\left(t'\right) P_j\left(t'-t\right) - \overline{P_i}\ \overline{P_j}\right\rangle_{t'}}{\sigma_i \sigma_j}, \tag{3.32}$$

where $\overline{P_i}$, $\overline{P_j}$ and σ_i, σ_j are the mean and standard deviations of the laser powers, respectively. Complete synchronization is quantitatively characterized by the mean synchronization error

$$\langle e \rangle = P_i - P_j. \tag{3.33}$$

Figure 3.32 shows C and $\langle e \rangle$ when Equations 3.32 and (3.33) are used; C_M, $\langle e_M \rangle$ are calculated for ML_R and ML_T (Figure 3.32(a)), C_T, $\langle e_T \rangle$ for ML_T and SL_T, C_R, $\langle e_R \rangle$ for ML_R and SL_R (Figure 3.32(b)), and C_S, $\langle e_S \rangle$ for SL_R and SL_T (Figure 3.32(c)) as functions of the coupling strengths for unidirectional and bidirectional coupling between ML_T and ML_R. The qualitative behavior in the two cases is similar, but a much smaller coupling is required to obtain complete synchronization between the masters (Figure 3.32(a)) and between the slaves (Figure 3.32(c)) for bidirectional coupling.

Another advantage of bidirectional coupling (of importance for security purposes) is that the transmitter has feedback information about the receiver's behavior. If an eavesdropper enters the synchronization channel to obtain part of (or all) the laser light, the feedback signal is modified. In fact, a hacker connected

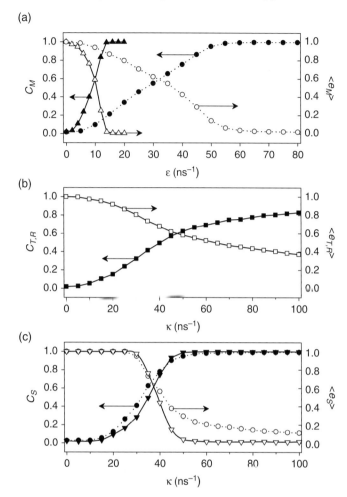

Figure 3.32 Cross-correlation (filled symbols) and mean synchronization error (empty symbols) between (a) ML_T and ML_R, (b) ML_T and SL_T, and (c) SL_T and SL_R, when ML_T and ML_R are coupled unidirectionally (circles) and bidirectionally (triangles) as functions of their coupling parameter. All cross-correlations are calculated in the absence of a message (reprinted with permission from Pisarchik and Ruiz-Oliveras 2010).

to the synchronization channel would have to use part of the laser power to synchronize his own laser, thus reducing the power entering the authorized receiver and, in turn, the power returning to the transmitter.

Since the master and slave lasers are not identical, generalized synchronization is characterized by a relatively small cross-correlation ($C \approx 0.8$) and a high mean error ($\langle e \rangle \approx 0.4$) even for very strong coupling ($\kappa = 100$ ns^{-1}) (Figure 3.32(b)). This indicates a large difference between the laser dynamics, preventing a nonauthorized person, even with access to both channels, from reading the message without knowing the generalized synchronization function between master and slave pairs. From Figures 3.32(b) and 3.32(c) one can see that complete synchronization between the master and slave lasers is not needed to obtain complete synchronization between SL_T and SL_R, and even a relatively small coupling strength ($\kappa \approx 50 \text{ ns}^{-1}$) produces an excellent correlation.

The ability of this system to provide high-quality communication has been demonstrated through numerical simulations of the Lang–Kobayashi model (Pisarchik and Ruiz-Oliveras 2010). Notably, bidirectional coupling between the master lasers provides much better communication quality (bandwidth of up to 5 Gb/s) than unidirectional coupling and makes the system more robust against a mismatch between the master laser parameters and to differences in the coupling strengths between master and slave pairs.

Communication using chaos synchronization remains a hot research topic, with many new articles and patents appearing every year. Therefore, one can be confident that future trends in this direction will be focused on the development of new systems based on different types of synchronization, on the implementation of laser networks, and on the prominent use of multistability and intermittency.

4

High-Dimensional Systems

The emergence of synchronization and its general intriguing features are of particular interest in systems with a large number (or even infinitely many) degrees of freedom.

From neurons to population dynamics and chemical waves, nature showcases many examples of synchronized and collective behavior in such high-dimensional systems. Even though many of the key principles have been established by the pioneering work of Winfree, Kuramoto, and others in the sixties and seventies of the last century, a full understanding of high-dimensional synchronization and self-organization is still missing.

The aim of this chapter is not to give a full account of the many subareas where synchronization in high-dimensional systems has been explored so far. Instead, our focus will be on a few such systems that have gained prominence in research over the recent years. A paradigmatic high-dimensional model is the Kuramoto model (Kuramoto 1975, 1984), which we have already used as a motivating example in Chapter 2. We will return to this model, explain some general details, and then explore more recently discovered phenomena such as Chimera and Bellerophon states, time-delayed auto-synchronization and amplitude death.

4.1 The Kuramoto Model

In order to be able to understand the fundamental mechanisms of the transition to synchronous behavior, it is desirable to use a model that is mathematically tractable and sufficiently generic. In 1975, Yoshiki Kuramoto published a three-page manuscript (Kuramoto, 1975) that introduced an elegant model that fulfils these requirements. This model, which is now known as the *Kuramoto Model*, has evolved into a cornerstone of our understanding of synchronization.

4.1.1 Derivation from a Generic Oscillator Model

Synchronization in natural systems often seems to have two essential "ingredients": (i) a number of individual systems that, in isolation, oscillate at different frequencies; and (ii) a coupling mechanism between these systems. Guided by this observation, Kuramoto considered a generic individual oscillator in the complex plane, which, when isolated, obeys the dynamical equation

$$\dot{Q} = (i\omega + \alpha)\, Q - \beta\, |Q|^2\, Q. \tag{4.1}$$

Here, ω is the natural frequency of the oscillator, α and $\beta > 0$ are parameters, and the dot denotes time derivative, as usual. Equation 4.1 passes through a supercritical Andronov–Hopf bifurcation at $\alpha = 0$, and for $\alpha > 0$ a stable limit cycle

$$Q^*(t) = \sqrt{\frac{\alpha}{\beta}}\, e^{i\omega t} \tag{4.2}$$

of period $T = 2\pi/\omega$ exists. Oscillators of this type are often called *Stuart–Landau* oscillators.

One would now like to couple an ensemble of N self-sustained oscillators Q_k with different natural frequencies ω_k, and in general there are many options to achieve this. Kuramoto proposed a particularly simple and natural coupling scheme, where all oscillators experience a common (identical) forcing term. This forcing term is taken to be proportional to the value of Q_k averaged over all oscillators. Written in formulas, the dynamical equations have the form

$$\dot{Q}_k = (i\omega_k + \alpha)\, Q_k - \beta\, |Q_k|^2\, Q_k + \frac{K}{N} \sum_{j=1}^{N} Q_j, \tag{4.3}$$

where K is a coupling strength parameter. The interaction term in Equation 4.3 corresponds to an *all-to-all* (or complete graph) connection topology, which sometimes is also called *mean-field* coupling. As far as the coupling is concerned, all oscillators are treated equally.

It is now useful to rewrite Equation 4.3 in terms of real dynamical variables $r_k(t)$ and $\vartheta_k(t)$ using the standard polar representation of complex numbers, i.e.,

$$Q_k = e^{i\vartheta_k} r_k. \tag{4.4}$$

Substituting this last expression into Equation 4.3 and dividing both sides by $e^{i\vartheta_k}$ we obtain

$$i\dot{\vartheta}_k r_k + \dot{r}_k = \left(i\omega_k + \alpha - \beta r_k^2 \right) r_k + \frac{K}{N} \sum_{j=1}^{N} r_j e^{i(\vartheta_j - \vartheta_k)}. \tag{4.5}$$

One can now split Equation 4.5 into real and imaginary parts, and obtain the dynamical equations

$$\dot{r}_k = \left(\alpha - \beta r_k^2\right) r_k + \frac{K}{N} \sum_{j=1}^{N} r_j \cos\left(\vartheta_j - \vartheta_k\right), \tag{4.6}$$

$$\dot{\vartheta}_k = \omega_k + \frac{K}{N} \sum_{j=1}^{N} \frac{r_j}{r_k} \sin\left(\vartheta_j - \vartheta_k\right). \tag{4.7}$$

Note that Equation 4.7 only holds for $r_k \neq 0$.

The dynamical system of Equations 4.6 and 4.7 is still too complex to allow for an easy analytical treatment. Kuramoto therefore decided to study it in the limit of infinitely large parameters $\alpha \to \infty$ and $\beta \to \infty$, while keeping the ratio α/β constant. The parameters α and β only appear in the dynamics for the radial variables in Equation 4.6, and we approximate this equation for large α and β as

$$\frac{\dot{r}_k}{\beta} = \left(\frac{\alpha}{\beta} - r_k^2\right) r_k + \frac{K}{\beta N} \sum_{j=1}^{N} r_j \cos\left(\vartheta_j - \vartheta_k\right) \approx \left(\frac{\alpha}{\beta} - r_k^2\right) r_k. \tag{4.8}$$

In this limit, the dynamics for each variable r_k fully decouple from the rest of the system, and $r_k(t)$ approaches the stable fixed point $r_k^* = \sqrt{\frac{\alpha}{\beta}}$ for every k. Furthermore, the characteristic time required to approach this stable fixed point scales with β^{-1} and thus becomes arbitrarily small in the limit of $\beta \to \infty$. We can therefore set $r_k(t) = r_k^*$ in Equation 4.7 and finally reach the *Kuramoto model*

$$\dot{\vartheta}_k = \omega_k + \frac{K}{N} \sum_{j=1}^{N} \sin\left(\vartheta_j - \vartheta_k\right). \tag{4.9}$$

The Kuramoto model is the simplest model that is able to describe the transition to synchronous behavior in an ensemble of nonidentical oscillators. It therefore serves as a starting point for the discussion of synchronization in high-dimensional systems, and a familiarity with the basic features of the Kuramoto model is a prerequisite for the understanding of synchronization in more complex circumstances.

Formally, the Kuramoto model in Equation 4.9 has N real degrees of freedom, namely the angular variables ϑ_k for $k = 1, \ldots, N$. The phase space is therefore an N-dimensional torus $\mathbb{T}^N = [0, 2\pi]^N$. On this torus, we can define the phase sum variable

$$\vartheta_\Sigma = \sum_{k=1}^{N} \vartheta_k, \tag{4.10}$$

for which it is

$$\dot{\vartheta}_\Sigma = \sum_{k=1}^{N} \omega_k = N\bar{\omega}, \tag{4.11}$$

where $\bar{\omega}$ is the average frequency.

Since only phase differences appear on the right hand side of Equation 4.9, the model is invariant under constant phase shifts. Explicitly, for a given solution $(\vartheta_1(t), \ldots, \vartheta_N(t))$ of Equation 4.9, the shifted vector $(\vartheta_1(t) + \alpha, \ldots, \vartheta_N(t) + \alpha)$ is again a solution for any constant α.

Similarly, it is also noted that a common constant shift of all frequencies does not change the dynamics of Equation 4.9 in a material way. Specifically, if $(\vartheta_1(t), \ldots, \vartheta_N(t))$ is a solution of Equation 4.9 for a given set of frequencies $(\omega_1, \ldots, \omega_N)$, then $(\vartheta_1(t) - \bar{\omega}t, \ldots, \vartheta_N(t) - \bar{\omega}t)$ is a solution of Equation 4.9 with shifted frequencies $(\omega_1 - \bar{\omega}, \ldots, \omega_N - \bar{\omega})$. It is therefore without loss of generality that we restrict ourselves to the study of systems where the average frequency vanishes, i.e., $\bar{\omega} = 0$. In this case, it follows from Equation 4.11 that ϑ_Σ is constant in time.

4.1.2 The Case $N = 3$

In Section 2.3 we have discussed the case of $N = 2$, and shown how to treat it. The case $N = 3$ is the first nontrivial case, and it merits a closer examination, since many of its features are also relevant for the transition to synchrony in general systems, even though it can still be solved by standard techniques of bifurcation theory (Maistrenko et al. 2004, 2005).

Without loss of generality, let us assume that the average frequency vanishes, and that $\vartheta_\Sigma = 0$. We can then write the system in Equation 4.9 as

$$\begin{aligned}
\dot{\vartheta}_1 &= \omega_1 + \frac{K}{3}[\sin(\vartheta_2 - \vartheta_1) + \sin(\vartheta_3 - \vartheta_1)] \\
\dot{\vartheta}_2 &= \omega_2 + \frac{K}{3}[\sin(\vartheta_1 - \vartheta_2) + \sin(\vartheta_3 - \vartheta_2)] \\
\dot{\vartheta}_3 &= -\vartheta_2 - \vartheta_1
\end{aligned} \tag{4.12}$$

This defines a flow on the standard torus $\mathbb{T}^2 = [0, 2\pi]^2$.

Let us now define the time-averaged frequencies for each individual oscillator as

$$\bar{\omega}_k = \lim_{t \to \infty} \frac{1}{t} \int_0^t \dot{\vartheta}_k(t') \, dt', \tag{4.13}$$

and study how they change as a function of the coupling strength K at fixed parameters ω_1 and ω_2.

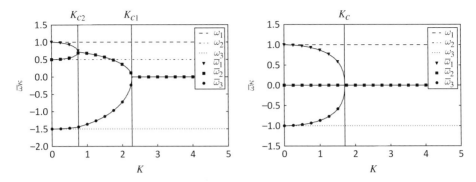

Figure 4.1 Bifurcation diagram of the average frequencies $\bar{\omega}_k$ versus coupling strength K for the Kuramoto model Equation 4.21 with $N = 3$. Parameters are $\omega_1 = 1$, $\omega_2 = 0.5$ and $\omega_3 = -1.5$ in the left-hand panel and $\omega_1 = 1$, $\omega_2 = 0$ and $\omega_3 = -1$ in the right-hand panel.

For $\omega_1 = 1$ and $\omega_2 = 0.5$ the corresponding bifurcation diagram is shown in the left-hand panel of Figure 4.1. It is observed that, at large $K > K_{c1}$, the system is in a phase-synchronized state, where all oscillators are phase locked, and thus share a common average (null) frequency. This situation is also illustrated in the first panel of Figure 4.2, where the phase-locked state is indicated by the three fixed points on the torus. The appearance of three fixed points, rather than just a single fixed point might be surprising. However, the reason for this threefold degeneracy is easily found by noting that the system 4.12 is invariant under the symmetry operation

$$s_3 : (\vartheta_1, \vartheta_2) \mapsto \left(\vartheta_1 + \frac{2\pi}{3}, \vartheta_2 + \frac{2\pi}{3} \right). \tag{4.14}$$

Since $(s_3)^3 = 1$ is the identity operation, the system is invariant under the symmetry group $\mathbb{Z}_3 = \{1, s_3, (s_3)^2\}$. As a consequence, every fixed point of the system is threefold degenerate.

As one decreases K below the critical value K_{c1} the system splits up into a phase-locked "cluster" of two oscillators, which share a common average frequency $\bar{\omega}_1 = \bar{\omega}_2 > 0$ and a third oscillator with a negative frequency $\bar{\omega}_3 = -2\bar{\omega}_1 < 0$. The mechanism for this transition is a saddle-node bifurcation, which is analogous to the unlocking transition discussed in Section 2.3.

The corresponding dynamics on the torus (close to the bifurcation point) are shown in the second panel of Figure 4.2. The dynamics now follow a limit cycle that visits all three regions where the fixed points had existed for $K > K_{c1}$. Note that this limit cycle is invariant under the symmetry operation s_3, and therefore only one (and not three) limit cycle exists.

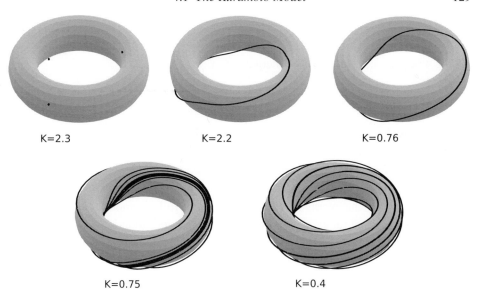

K=2.3   K=2.2   K=0.76

K=0.75   K=0.4

Figure 4.2 Dynamics of the Kuramoto model for $N = 3$ projected onto the torus $(\vartheta_1, \vartheta_2)$ for different values of K. The other parameters are $\omega_1 = 1$, $\omega_2 = 0.5$ and $\omega_3 = -1.5$.

The value of K_{c1} is not known analytically. However, if for a particular value of K one can show that the condition $\dot{\vartheta}_k \neq 0$ is fulfilled for at least one of the oscillators k, then it follows that the system has no fixed points. Since we know from Equation 4.12 that

$$\left| \dot{\vartheta}_k \right| \geqslant |\omega_k| - \frac{2K}{3} ,$$ (4.15)

this yields the simple lower bound

$$K_{c1} > \frac{3}{2} \max_{k \in \{1,2,3\}} |\omega_k| .$$ (4.16)

For the example with $\omega_1 = 1$ and $\omega_2 = 0.5$, we obtain numerically $K_{c1} \approx 2.282$ and $3 \left| \omega_3 \right| / 2 = 2.25$, which confirms the inequality.

For even smaller K, below a critical value K_{c2} the cluster splits, and we obtain three different average frequencies $\bar{\omega}_1 \neq \bar{\omega}_2 \neq \bar{\omega}_3$. This transition to a fully desynchronized state is illustrated in panels 3 and 4 of Figure 4.2. In panel 3, we see the same limit cycle that was obtained at K_{c1} and that forms a closed loop after one rotation of both ϑ_1 and ϑ_2. However, in panel four, the limit cycle has "spread out" over the torus. After n rotations in ϑ_1 now only $n - 1$ rotations in ϑ_2 are obtained, where n diverges to infinity as K approaches K_{c2} from below. We can find a lower bound for K_{c2} by the requirement that $\left| \dot{\vartheta}_k - \dot{\vartheta}_j \right| > 0$ for all k and j. This yields

$$\left|\dot{\vartheta}_k - \dot{\vartheta}_j\right| \geqslant \left|\omega_k - \omega_j\right| - \frac{2K}{3}. \tag{4.17}$$

As a consequence, we obtain the inequality

$$K_{c2} > \frac{3}{2} \min_{k,j \in \{1,2,3\}} \left|\omega_k - \omega_j\right|. \tag{4.18}$$

For the current example, we have $3(\omega_1 - \omega_2)/2 = 0.75$ and $K_{c2} \in (0.75, 0.76)$, confirming the inequality.

As one further decreases K towards 0, the average frequencies $\bar{\omega}_1$, $\bar{\omega}_2$, and $\bar{\omega}_3$ tend towards their respective free running frequencies ω_1, ω_2, and ω_3. While at the scale of Figure 4.1 the average frequencies seem to follow smooth curves, a closer inspection shows that they are actually piecewise-smooth functions with small vertical jumps. This is a consequence of the fact that there are infinitely many resonant limit cycles with n rotations of ϑ_1 and m rotations of ϑ_2, where n and m are integer numbers. One such resonant state is shown in the last panel of Figure 4.2.

So far we have studied the specific case of $\omega_1 = 1$ and $\omega_2 = 0.5$, because its two-step sequence is representative of a large parameter range. Another interesting scenario occurs when one of the ω_k is much smaller than the other two in absolute value. For example, for $\omega_1 = 1$ and $\omega_2 = 0$ the bifurcation diagram is shown in the right hand panel of Figure 4.1. In this case we observe that the system is again in a fully locked state for large K and in a fully unlocked state for small K, however, the transition between the two regimes now occurs in a single step at K_c, rather than in two separate steps as before.

The full bifurcation picture of the $N = 3$ Kuramoto model is actually much more complex, and may also contain regions of multistability between the fully locked and the cluster states. This is, for instance, the case for the bifurcation diagram shown in Figure 4.3, where one sees that the down sweep branch follows the fully locked state, even after the cluster state became stable. This leads to a small region of hysteresis as indicated by the arrows. Which of the two stable states is selected within the hysteresis region depends on the initial conditions. We stress, however, that these phenomena occur within rather small parameter intervals, and, on a larger scale, the scenario in Figure 4.3 can be overlooked. For further details on the bifurcations of the $N = 3$ Kuramoto model we refer the interested reader to the main results available in the literature (Maistrenko et al. 2004).

4.1.3 The Kuramoto Order Parameter

In the previous section we have described some of the interesting phenomena occurring in the Kuramoto model for $N = 3$. Similar phenomena have been observed for other small values of N (Chiba and Pazó, 2009). For $N \geqslant 4$ chaotic behavior is also found (Maistrenko et al. 2005; Popovych et al. 2005). Rather than

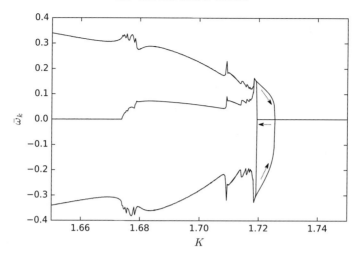

Figure 4.3 Bifurcation diagram for the $N = 3$ Kuramoto model with $\omega_1 = 1$, $\omega_2 = 0.01724$ and $\omega_3 = -\omega_1 - \omega_2$.

discussing these phenomena in detail, here we focus on the case of a large number of oscillators.

In order to assess and quantify synchrony in an ensemble of phase oscillators, it is common practice to introduce the two Kuramoto order parameters $r(t)$ and $\Psi(t)$ through

$$r(t)e^{i\Psi(t)} = \frac{1}{N}\sum_{j=1}^{N} e^{i\vartheta_j(t)}. \tag{4.19}$$

The function $r(t) \in [0, 1]$ provides a measure for synchrony: one obtains $r(t) = 1$ only in the case of complete synchronization of all oscillators, i.e., if $\vartheta_1(t) = \cdots = \vartheta_n(t)$.

The function $\Psi(t)$ is sometimes called the "average phase"; however, it should be stressed that in general Ψ is not equal to the arithmetic average of the angles ϑ_k (except in the case $N = 2$).

The order parameters defined in Equation 4.19 allow us to rewrite Equation 4.9 in the form

$$
\begin{aligned}
\dot{\vartheta}_k &= \omega_k + \frac{K}{N}\sum_{j=1}^{N}\Im\left(e^{i(\vartheta_j - \vartheta_k)}\right) \\
&= \omega_k + K\Im\left(\left(\frac{1}{N}\sum_{j=1}^{N}e^{i\vartheta_j}\right)e^{-i\vartheta_k}\right) \\
&= \omega_k + K\Im\left(re^{i(\Psi - \vartheta_k)}\right).
\end{aligned}
\tag{4.20}
$$

An alternative formulation of the Kuramoto model in Equation 4.9 is therefore given by the self-contained equations

$$\dot{\vartheta}_k = \omega_k + Kr \sin{(\Psi - \vartheta_k)} \, , \qquad (4.21)$$

$$re^{i\Psi} = \frac{1}{N} \sum_{j=1}^{N} e^{i\vartheta_j}. \qquad (4.22)$$

Apart from K and N, the only parameters in Equation 4.21 are the frequencies ω_k. Thus, for the model to be well defined, one must specify a rule to choose them. The classical choice is to take the ω_k as independent random variables with a specified probability density function $g(\omega)$ (Kuramoto 1975). Gaussian, uniform or Lorentzian distributions are popular choices for $g(\omega)$.

4.1.4 Numerical Phenomenology for Large N

For numerical simulations of the Kuramoto model at large N, the use of the formulation of Equation 4.21 is convenient, since the number of operations required to calculate its right-hand side scales only linearly with N. This compares very favorably with the original formulation 4.9, where the number of operations scales with N^2. Other than that, the numerical implementation is straightforward, and a standard Runge–Kutta scheme for the integration of ordinary differential equations can be employed.

As an example, let us study the case of $N = 2048$ with a normal frequency distribution given by

$$g(\omega) = \frac{1}{\sqrt{2\pi}} e^{-\frac{\omega^2}{2}}. \qquad (4.23)$$

The initial phases are chosen to be randomly and uniformly distributed over the interval $[0, 2\pi]$.

Figure 4.4 shows the corresponding time traces of $r(t)$ for a number of different coupling strengths K. In addition, Figure 4.5 depicts snapshots of the resulting phases at $t = 200$. The general qualitative trend that emerges from these simulations is that for $K \leqslant 1$ the order parameter $r(t)$ remains small (less than approximately 0.1) at all times and fluctuates in an aperiodic fashion. In this case the coupling between oscillators is not sufficient to overcome the differences in local frequencies and, as a consequence, the phases in the lower-right panel of Figure 4.5 are uniformly distributed over the unit circle.

For $K \geqslant 2$, the curves of $r(t)$ in Figure 4.4 fluctuate around well-defined long-time average values. The interpretation of this result is that for large K a major part of the oscillators phase-synchronize. This is also supported by the corresponding

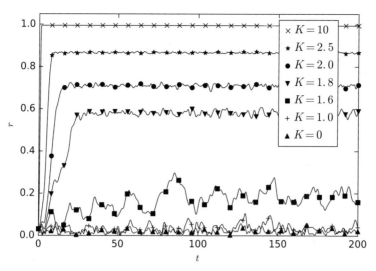

Figure 4.4 Numerical time traces of the order parameter $r(t)$ of the Kuramoto model (Equation 4.21) for a normal distribution of $g(\omega)$ and $N = 2,048$, for different values of K. The initial phases are randomly and uniformly chosen in the interval $[0, 2\pi]$.

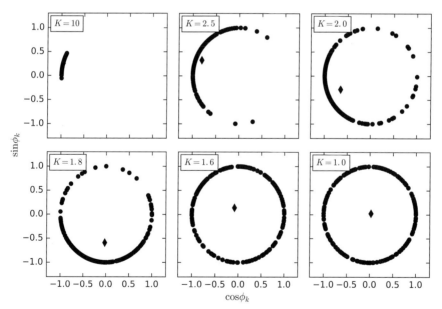

Figure 4.5 Snapshots of $e^{i\vartheta_k}$ at $t = 200$ for the Kuramoto model. To increase clarity, only 10 percent of the ϑ_k values are plotted as dots on the unit circle. The diamonds indicate the complex order parameter $re^{i\Psi}$ of Equation 4.19. Values of K are indicated in the subplot titles. All the other parameters are set as in Figure 4.4.

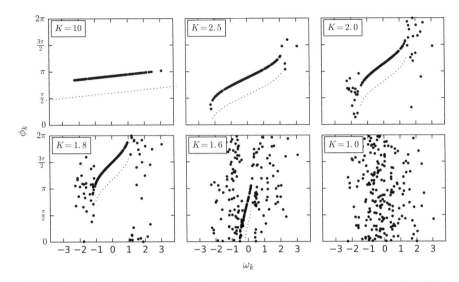

Figure 4.6 Snapshots of ϑ_k at $t = 200$ versus ω_k for the Kuramoto model. The dashed line shows the position of the stable fixed point $\vartheta_k^* = \varphi_k^* + \Psi$, where φ_k^* is obtained from Equation 4.38. The dashed line has been shifted vertically by a constant value of 1, to increase clarity. Numerical parameters are as in Figure 4.4.

phase snapshots in the top row of Figure 4.5, which show that most phases cluster around a single value, with a spread that decreases with K.

While the cases of very large and very small K are easily explained in qualitative terms, the most interesting regime is that of intermediate K, where the actual transition to synchronization occurs. Some insight into this can be obtained from Figure 4.6, which is similar to Figure 4.5, but shows the phases ϑ_k versus ω_k. For instance, one can observe that for $K = 1.6$ a distinct phase-synchronized cluster with $\omega_k \approx 0$ is formed. In this cluster, the phases gradually increase with ω_k. Outside the cluster, the dynamics of the phases are unsynchronized.

Comparing the panel for $K = 1.6$ with the others, one sees that this cluster is completely absent for $K = 1$, while it has considerably increased at $K = 1.8$. For $K = 2.5$ almost all oscillators are part of the synchronization cluster, with only a few oscillators at the fringes not participating in the phase-synchronized dynamics. At $K = 10$, the synchronization cluster encompasses all oscillators and the system is fully phase-synchronized. Our numerical results suggest therefore that the mechanism of the transition to synchrony involves the birth and growth of a synchronization cluster that starts at the central frequency of the frequency distribution $g(\omega)$.

While providing useful insights into the Kuramoto model, it should be emphasized that numerical results are subject to large variations in the transition region.

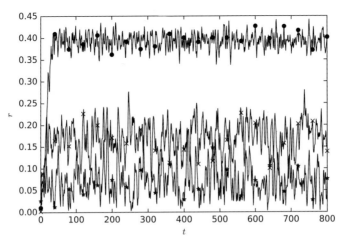

Figure 4.7 Three numerical time traces of the order parameter $r(t)$ of the Kuramoto model for $K = 1.6$. All other parameters are as in Figure 4.4.

To illustrate the difficulties that one may encounter in this regime, let us consider three different time traces of $r(t)$ for $K = 1.6$ (Figure 4.7). One sees that the three runs lead to very different results for the long-time average of $r(t)$, although they were performed with the same parameters. The reason for these differences is that for each run a new set of random frequencies ω_k and initial conditions $\vartheta_k(0)$ was chosen. In particular, the closeness of the random frequencies ω_k to the central zero frequency has a critical influence on the existence (and the size) of the synchronization cluster.

4.1.5 Theory for Large N

As we have started seeing, the Kuramoto model is deceptively simple, and despite decades of active research it is still not completely understood. The current state of the art is regularly reviewed in the literature (Acebrón et al. 2005; Chiba 2015; Rodrigues et al. 2016), and we do not intend to give a complete and mathematically rigorous overview of the available results. Instead, we aim at using the Kuramoto model as a prototype system, which allows us to introduce a number of key concepts that are useful in broader contexts.

Derivation of a Continuous Limit

Above, we have shown that the Kuramoto model as formulated in Equations 4.21 and 4.22 for finite N depends on the set of frequencies $\Omega = \{\omega_1, \dots, \omega_N\}$, usually chosen according to a given probability density $g(\omega)$. Thus, all individual realizations of the model are different. In the limit of $N \to \infty$ the set of selected frequencies becomes countably infinite, and dense on the support of $g(\omega)$ with probability 1.

The idea is then to introduce an oscillator density functional $\rho_t(\vartheta, \omega)$, which represents the density of oscillators per angle ϑ at given frequency ω and time t. As a consequence, the normalization condition

$$\int_0^{2\pi} \rho_t(\vartheta, \omega)\, d\vartheta = 1 \tag{4.24}$$

holds for all t and ω. Using $\rho_t(\vartheta, \omega)$, the Kuramoto order parameter of Equation 4.19 now becomes

$$r(t)\, e^{i\Psi(t)} = \int_{-\infty}^{\infty} \int_0^{2\pi} g(\omega)\, \rho_t(\vartheta, \omega)\, e^{i\vartheta}\, d\vartheta\, d\omega. \tag{4.25}$$

In a similar way, one can now rewrite Equation 4.21 as

$$\dot{\vartheta}(\vartheta, \omega) = \omega + K r \sin(\Psi - \vartheta), \tag{4.26}$$

which represents the dynamics of individual oscillators at a given frequency and angle.

This allows one to derive a dynamical equation for $\rho_t(\vartheta, \omega)$. For a given frequency ω and time t, The fraction q_t of oscillators in some small angle interval $[\vartheta, \vartheta + \delta\vartheta]$ is

$$q_t = \int_{\vartheta}^{\vartheta + \delta\vartheta} \rho_t(\vartheta', \omega)\, d\vartheta' \approx \rho_t(\vartheta, \omega)\, \delta\vartheta. \tag{4.27}$$

Within a small time interval δt, this fraction q_t changes due to oscillators crossing the boundaries either at ϑ or at $\vartheta + \delta\vartheta$. For example, in the case of $\dot{\vartheta}(\vartheta, \omega) > 0$, oscillators are *entering* the interval at ϑ and contribute to an increase of q_t. In the first order of δt, the fraction of oscillators entering at ϑ is proportional to both the angular velocity and the density of oscillators at this boundary, and is explicitly given by $\rho_t(\vartheta, \omega)\, \dot{\vartheta}(\vartheta, \omega)\, \delta t$.

Similarly, the fraction of oscillators *leaving* at $\vartheta + \delta\vartheta$ is given by $\rho_t(\vartheta + \delta\vartheta, \omega)$ $\dot{\vartheta}(\vartheta + \delta\vartheta, \omega)\, \delta t$. Specular considerations apply for negative $\dot{\vartheta}$. In the first order of δt, the total change in q_t can be expressed as

$$q_{t+\delta t} \approx q_t + \rho_t(\vartheta, \omega)\, \dot{\vartheta}(\vartheta, \omega)\, \delta t - \rho_t(\vartheta + \delta\vartheta, \omega)\, \dot{\vartheta}(\vartheta + \delta\vartheta, \omega)\, \delta t. \tag{4.28}$$

Dividing by both δt and $\delta\vartheta$, using Equation 4.27 and taking the limit of $\delta t, \delta\vartheta \to 0$, we get

$$\dot{\rho}_t(\vartheta, \omega) = -\partial_\vartheta \left(\rho_t(\vartheta, \omega)\, \dot{\vartheta}(\vartheta, \omega) \right). \tag{4.29}$$

It is therefore possible to replace the system of ordinary differential equations 4.21 and 4.22 for the dynamical variables ϑ_k with a partial differential equation system for $\rho_t(\vartheta, \omega)$ given by

$$\dot{\rho}_t \left(\vartheta, \omega\right) = -\partial_\vartheta \left(\rho_t \left(\vartheta, \omega\right) \left[\omega + Kr \sin\left(\Psi - \vartheta\right)\right]\right),$$

$$r(t)\, e^{i\Psi(t)} = \int_{-\infty}^{\infty} \int_0^{2\pi} g\left(\omega\right) \rho_t \left(\vartheta, \omega\right) e^{i\vartheta}\, d\vartheta\, d\omega. \tag{4.30}$$

It should be stressed that our "derivation" of Equation 4.30 from the finite-N Kuramoto model does not satisfy the requirements of strict mathematical rigor. In particular, we have not discussed the function spaces in which we are to find the solution $\rho_t \left(\vartheta, \omega\right)$, and we avoided discussing in which probabilistic sense solutions of the discrete and continuous Kuramoto models are related.

The general solution of Equation 4.30 is quite complex. However, a simple solution always exists regardless of the choice of $g\left(\omega\right)$ and K, and it is the fully desynchronized solution described by

$$\rho_t \left(\vartheta, \omega\right) = \frac{1}{2\pi} \tag{4.31}$$

for all t and ω. In this case, one finds that $r = 0$ and Ψ is undefined. All oscillators follow their natural frequency and the oscillator density is evenly distributed over all angles.

The numerical results of Section 4.1.4 suggest that this solution might be stable for small K, but should become unstable for sufficiently large K. Unfortunately, the stability analysis of the system of Equation 4.30 is difficult, and involves the study of the continuous spectrum of infinite-dimensional operators. Mathematically, satisfactory treatments are available in the literature, but they require some background in functional analysis for their understanding. We refer the interested reader to the classical treatment by Strogatz and Mirollo (1991), and to a recent work of Chiba (2015), which uses the spectral theory of rigged Hilbert spaces to derive the stability properties of the unsynchronized solution.

The Kuramoto Transition

Instead of discussing the stability of the unsynchronized solution of Equation 4.31, an alternative approach is to try and find other simple solutions of the system in Equation 4.30. Following the treatment of Kuramoto (1984), one seeks solutions where the Kuramoto order parameters are given by the expressions

$$r(t) = \hat{r},$$
$$\Psi(t) = \hat{\omega}t, \tag{4.32}$$

where $\hat{r} > 0$ and $\hat{\omega}$ are two constants, to be determined. For such a solution (if it exists), the dynamics at each individual frequency ω can be obtained from the Adler equation 2.7. As in Section 2.3.1, one introduces a shifted coordinate frame $\varphi = \vartheta - \Psi$. For the transformed density functional one has

$$\hat{\rho}_t \left(\varphi, \omega\right) = \rho_t \left(\varphi + \Psi, \omega\right). \tag{4.33}$$

The system in Equation 4.30 then takes the form

$$\frac{d}{dt}\hat{\rho}_t\left(\varphi,\omega\right) = -\partial_\varphi\left(\hat{\rho}_t\left(\varphi,\omega\right)\left[\omega - \hat{\omega} - K\hat{r}\sin\left(\varphi\right)\right]\right), \tag{4.34}$$

$$\hat{r} = \int_{-\infty}^{\infty} g\left(\omega\right)\int_0^{2\pi}\hat{\rho}_t\left(\varphi,\omega\right)e^{i\varphi}d\varphi d\omega. \tag{4.35}$$

We now want to find a stationary solution of Equation 4.34, which does not depend on time. This leads to the following condition:

$$\hat{\rho}_t\left(\varphi,\omega\right)\left[\omega - \hat{\omega} - K\hat{r}\sin\left(\varphi\right)\right] = C\left(\omega\right), \tag{4.36}$$

where $C\left(\omega\right)$ is independent of φ. There are only two cases in which Equation 4.36 is satisfied, depending on the roots of the expression in square brackets. If the expression can never vanish, which means that $\left|\omega - \hat{\omega}\right| > K\hat{r}$, the dynamics are unlocked. In this case, from the normalization condition for $\hat{\rho}_t\left(\varphi,\omega\right)$, it follows that

$$\hat{\rho}_t\left(\varphi,\omega\right) = \frac{\sqrt{\left(\omega - \hat{\omega}\right)^2 - \left(K\hat{r}\right)^2}}{2\pi\left|\omega - \hat{\omega} - K\hat{r}\sin\left(\varphi\right)\right|}. \tag{4.37}$$

If, instead, $\left|\omega - \hat{\omega}\right| \leqslant K\hat{r}$, then there are one or two values of φ for which the expression in square brackets vanishes. In this case, corresponding to locked dynamics, $\hat{\rho}_t\left(\varphi,\omega\right)$ cannot be normalized for $C\left(\omega\right) \neq 0$, and therefore $C\left(\omega\right)$ must vanish as well. This means that the weight of $\hat{\rho}_t\left(\varphi,\omega\right)$ must be concentrated at the zeros of $\left[\omega - \hat{\omega} - K\hat{r}\sin\left(\varphi\right)\right]$, i.e., either at

$$\psi_s^* = \arcsin\frac{\omega - \hat{\omega}}{K\hat{r}} \tag{4.38}$$

or at

$$\varphi_u^* = \pi - \varphi_s^*. \tag{4.39}$$

From our previous discussion of the Adler equation, we know that φ_u^* corresponds to an unstable fixed point. Thus, we conclude that in a stable situation the entire weight of the distribution is concentrated at φ_s^*, and it has the form

$$\hat{\rho}_t\left(\varphi,\omega\right) = \delta\left(\varphi - \varphi_s^*\right). \tag{4.40}$$

To summarize, the stable stationary probability distribution for Equation 4.34 is given by

$$\hat{\rho}_t\left(\varphi,\omega\right) = \begin{cases} \frac{\sqrt{\left(\omega - \hat{\omega}\right)^2 - \left(K\hat{r}\right)^2}}{2\pi\left|\omega - \hat{\omega} - K\hat{r}\sin(\varphi)\right|} & \text{for } \left|\omega - \hat{\omega}\right| > K\hat{r}, \\ \delta\left(\varphi - \arcsin\frac{\omega - \hat{\omega}}{K\hat{r}}\right) & \text{otherwise.} \end{cases} \tag{4.41}$$

Notice that Equation 4.41 has been derived from Equation 4.34 alone. However, Equation 4.35 also needs to be fulfilled. This latter equation contains a double integral over ω and φ. It is useful to first integrate over φ and consider the respective contributions from the locked and unlocked parts of the system separately.

For the contribution of the locked oscillators, one finds

$$
\int_0^{2\pi} \hat{\rho}_t\,(\varphi, \omega)\,e^{i\varphi}\,d\varphi = \int_0^{2\pi} \delta\left(\varphi - \arcsin\frac{\omega - \hat{\omega}}{K\hat{r}}\right) e^{i\varphi}\,d\varphi
$$

$$
= e^{i\arcsin\frac{\omega - \hat{\omega}}{K\hat{r}}} \tag{4.42}
$$

$$
= i\frac{\omega - \hat{\omega}}{K\hat{r}} + \sqrt{1 - \left(\frac{\omega - \hat{\omega}}{K\hat{r}}\right)^2}.
$$

In the unlocked case, one needs to calculate the integral over the probability distribution, which is of the form

$$
\int_0^{2\pi} \hat{\rho}_t\,(\varphi, \omega)\,e^{i\varphi}\,d\varphi = \frac{\sqrt{a^2 - 1}}{2\pi} \int_0^{2\pi} \frac{e^{i\varphi}\,d\varphi}{|a - \sin(\varphi)|}, \tag{4.43}
$$

where we put

$$
a = \frac{\omega - \hat{\omega}}{K\hat{r}}. \tag{4.44}
$$

This can be integrated using the substitution $z = e^{i\varphi}$ and the residue theorem. For $a > 1$, one obtains

$$
\int_0^{2\pi} \frac{e^{i\varphi}}{a - \sin\varphi}\,d\varphi = \frac{1}{i}\int_{|z|=1} \frac{dz}{a - \frac{1}{2i}\left(z - z^{-1}\right)}
$$

$$
= \int_{|z|=1} \frac{-2z\,dz}{z^2 - 2iaz - 1}
$$

$$
= \int_{|z|=1} \frac{-2z\,dz}{(z - z_+)\,(z - z_-)} \tag{4.45}
$$

$$
= 2\pi i\frac{-2z_-}{z_- - z_+}
$$

$$
= 2\pi i\frac{\left(a - \sqrt{a^2 - 1}\right)}{\sqrt{a^2 - 1}},
$$

where $z_\pm = i\left(a \pm \sqrt{a^2 - 1}\right)$. Substitution into Equation 4.43 yields

$$
\int_0^{2\pi} \hat{\rho}_t\,(\varphi, \omega)\,e^{i\varphi}\,d\varphi = i\left[\frac{\omega - \hat{\omega}}{K\hat{r}} \mp \sqrt{\left(\frac{\omega - \hat{\omega}}{K\hat{r}}\right)^2 - 1}\right]. \tag{4.46}
$$

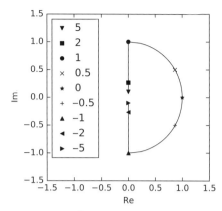

Figure 4.8 Complex values of $\int_0^{2\pi} \hat{\rho}_t(\varphi, \omega) e^{i\varphi} d\varphi$ according to Equations 4.42 and 4.46 for different values of a.

In this last formula, the positive sign before the square root applies for the case of $\omega - \hat{\omega} < -K\hat{r}$. We observe that the contribution from the unlocked oscillators to the Kuramoto order parameter is purely imaginary.

The contributions from Equations 4.42 and 4.46 to the Kuramoto order parameter for different values of a are shown in Figure 4.8. The contributions from the locked oscillators fall onto the semicircle with positive real part, while the contributions from the unlocked oscillators fall onto the imaginary axis inside the unit circle.

Using Equations 4.42 and 4.46 one can now rewrite the self-consistency condition of Equation 4.35 in the form

$$1 = K \int_{-1}^{1} g\left(\hat{\omega} + aK\hat{r}\right) \sqrt{1 - a^2} da \,, \tag{4.47}$$

$$0 = \int_{-\infty}^{\infty} g\left(\hat{\omega} + aK\hat{r}\right) a\, da - \int_{1}^{\infty} g\left(\hat{\omega} + aK\hat{r}\right) \sqrt{a^2 - 1}\, da$$
$$+ \int_{-\infty}^{1} g\left(\hat{\omega} + aK\hat{r}\right) \sqrt{a^2 - 1}\, da \,. \tag{4.48}$$

Here Equation 4.47 is derived from the real part and Equation 4.48 is derived from the imaginary part of Equation 4.35. The two equations contain two unknown parameters $\hat{r}$ and $\hat{\omega}$.

In the classical case studied by Kuramoto, it is assumed that $g(\omega)$ is a unimodal even function. Then, Equation 4.48 is fulfilled for $\hat{\omega} = 0$. In this case, we can expand $g(\omega)$ about its unique maximum at 0,

$$g(\omega) \approx g(0) + \frac{g''(0)}{2}\omega^2. \tag{4.49}$$

This allows one to rewrite 4.47 as

$$1 = K g\left(0\right) \frac{\pi}{2} + \frac{g''\left(0\right)\pi}{16} K^3 \hat{r}^2. \tag{4.50}$$

Since $g''\left(0\right) < 0$, it follows that there is a critical value

$$K_c = \frac{2}{\pi g\left(0\right)}, \tag{4.51}$$

such that for $K < K_c$ Equation 4.50 has no real solution for $\hat{r}$. For K very close to K_c, the order parameter grows in the characteristic square-root form

$$\hat{r} \approx \frac{4}{K_c^2 \sqrt{-g''\left(0\right)\pi}} \sqrt{K - K_c} . \tag{4.52}$$

The emergence of a state with constant $\hat{r}$ is called the *Kuramoto transition*. We have not addressed the stability of this state, which is mathematically difficult. It has however been proved recently by Chiba (2015) that if $g\left(\omega\right)$ is either a rational, even, unimodal, and bounded function, or if it is Gaussian, then, under certain conditions on the initial values and the coupling strength K, the order parameter $\hat{r}$ approaches the value of 4.52 in linear order of $K - K_c$ in the limit of $t \to \infty$.

4.1.6 Kuramoto Model with Time-Varying Links

The Kuramoto model has been the basis for many similar models that have been studied later on in the literature. The details on how the original formulation should be adapted depend on the intended applications and modeling situations. In particular, the feature of homogeneous global coupling, i.e., the fact that in the original model all oscillators couple identically to all other oscillators, is often perceived to be unrealistic in practical situations.

As an interesting example, let us consider the model of Gutiérrez et al. (2011), which is motivated by the idea of Hebbian learning (Hebb 1949). Hebb introduced the idea that the strength of synaptic connections between two neurons in the brain should be modified according to their activity patterns. More specifically, if neuron A "repeatedly and persistently takes part in firing" neuron B, then "A's efficiency (as one of the cells firing B) is increased" (Hebb 1949).

To connect this paradigm with the Kuramoto model, in a crude approximation, each single neuron is taken to be a phase oscillator. The firing of the neuron is then associated with the crossing of the phase oscillator through a particular phase angle, which one can set to 0 for simplicity.

Without coupling, neurons are assumed to fire at random frequencies, in correspondence to the random frequency selection of the individual oscillators in the Kuramoto model. To translate the Hebbian learning rule into the language of phase oscillators, for any pair of oscillators ϑ_j and ϑ_k one considers a quantity $q_{k,j}$ that evolves according to the rule

$$T\dot{q}_{k,j} = -q_{k,j} + e^{i(\vartheta_k - \vartheta_j)}. \qquad (4.53)$$

Here T is a parameter that specifies the characteristic memory time. The long-term evolution of $q_{k,j}$ depends on the synchronization patterns of ϑ_j and ϑ_k. If the two oscillators are phase-synchronized with a fixed phase difference $\vartheta_k - \vartheta_j$, then $q_{k,j}$ converges to $e^{i(\vartheta_k - \vartheta_j)}$, which is on the unit circle of the complex plane. In all other cases, $q_{k,j}$ will remain strictly inside the unit circle. Note that $q_{k,j}$ can be explicitly expressed as

$$q_{k,j}(t) = \frac{1}{T} \int_{-\infty}^{t} dt' e^{-\frac{t-t'}{T} + i(\vartheta_k(t') - \vartheta_j(t'))}, \qquad (4.54)$$

and thus its modulus can be interpreted as a measure of synchronization between the participating oscillators that is averaged over the past history with an exponentially decaying weight function.

In the context of Hebbian learning, $|q_{k,j}|$ provides a measure of the extent by which oscillator k "takes part in firing" oscillator j. It should be noted, however, that this measure is symmetric, i.e., $|q_{k,j}| = |q_{j,k}|$, and therefore it does not provide information about causality.

Following the Hebbian learning rule, the connection strength $w_{k,j}$ from oscillator j to oscillator k is not static, but changes dynamically. Connections between synchronized pairs of oscillators are to be preferred over less synchronized connections. In addition, it is natural to assume that the total weight of all incoming connections is normalized to one, i.e., $\sum_j w_{k,j} = 1$ for all k. An evolution equation that fulfils these demands is given by

$$\dot{w}_{k,j} = |q_{k,j}| - \left(\sum_l |q_{k,l}| \right) w_{k,j}, \quad j, l \in \mathcal{N}_k. \qquad (4.55)$$

Here $\mathcal{N}_k$ is a set of randomly chosen neighbors of oscillator k, which reflects the idea that a given neuron only has a finite number of other neurons it can potentially connect to.

The weights obtained from Equation 4.55 are then used in a Kuramoto-type equation of the form

$$\dot{\vartheta}_k = \omega_k + \lambda \sum_{j \in \mathcal{N}_k} w_{k,j} \sin\left(\vartheta_j - \vartheta_k\right), \qquad (4.56)$$

where the frequencies ω_k are randomly chosen and λ is a coupling parameter.

The system given by Equations 4.55 and 4.56 was investigated by Gutiérrez et al. (2011), and the key result is reprinted in Figure 4.9. In the left-hand column, histograms for the link weights $w_{k,j}$ are shown for various values of the coupling

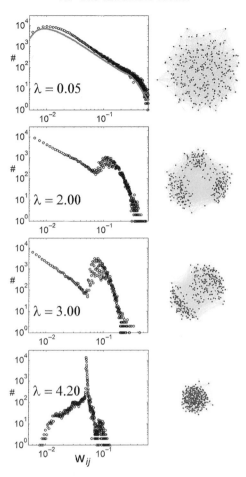

Figure 4.9 Left column: weight distribution for the Kuramoto model with dynamical link weights given by Equations 4.56 and 4.55 for different values of coupling strength λ. Parameters are $N = 300$, $\#\mathcal{N}_k = 20$ and ω_k uniformly distributed in $[-\pi, \pi]$. Right column: sketch of the corresponding network topology, where link lengths are inversely proportional to their weights. The solid line in the top-left panel represents the analytically derived weight distribution for $\lambda = 0$ (see Equation (4) in Gutiérrez et al. 2011). Reprinted with permission from Gutiérrez et al. (2011).

parameter λ. It is observed that for small $\lambda = 0.05$, the distribution of weights closely follows the distribution at $\lambda = 0$, which can be obtained analytically (solid line in top-left panel).

Over a large range of link weights, the distribution follows a power law distribution $p\left(w_{k,j}\right) \sim w_{k,j}^{\gamma}$ as indicated by the region of negative constant slope in the bilogarithmic histogram in the top left panel of Figure 4.9. Distributions of links that follow a power-law are called "scale-free" and are heuristically encountered in many natural systems. The dynamical link strength model can therefore provide a possible explanation for the emergence of scale-free distributions in an ensemble of coupled oscillators. Another explanation for scale-free weight distributions using evolutionary networks will be explored in Chapter 5.

The top-right panel of Figure 4.9 indicates that the system is in the unsynchronized state, as expected for small coupling. With increasing λ ($\lambda = 2$ and $\lambda = 3$ in Figure 4.9) the dynamics show a transition into a number of competing phase synchronized clusters of roughly equal size. Intracluster links are enhanced compared to intercluster links, and the corresponding histograms develop a characteristic "hump," roughly located at $w_{k,j} = \#\text{clusters}/\#\mathcal{N}_k$.

For very large $\lambda = 4.2$, the system evolves into a state with only one synchronization cluster, with the majority of oscillators participating in a phase-locked motion. Almost all links have now identical weights $w_{k,j} = 1/\#\mathcal{N}_k$, and the overall dynamics are therefore again similar to the standard Kuramoto model.

4.2 High-Dimensional Systems with Spatial Topologies

In the previous section, we introduced the Kuramoto model as a paradigmatic model for synchronization in high-dimensional systems and provided some examples of its usefulness. As all oscillators are identically coupled, they include no notion of spatial location. While this aspect of the model might be justified in many circumstances, there are equally important situations where spatial information is of crucial importance to explain observed phenomena. The main purpose of this section is to establish the terminology used to distinguish different coupling topologies in the context of spatial coupling. A more general formalism for coupling will be considered in Chapter 5 in the context of complex networks.

4.2.1 Spatially Discrete versus Spatially Continuous Systems

One common source for high-dimensional models are *spatially continuous* systems. A typical example is a model for the dynamics of a set of chemical species as a function of space and time. If one works in a d-dimensional space with m different species, one can introduce a function

$$\mathbf{u} : \mathbb{R}^d \times \mathbb{R} \to \mathbb{R}^m, \tag{4.57}$$

such that $\mathbf{u}(\mathbf{x}, t) \in \mathbb{R}^m$ denotes the concentration of the m substances at a point $\mathbf{x} \in \mathbb{R}^d$ and time t.

Very often, it is possible to restrict the study to some bounded region $V \subset \mathbb{R}^d$, and thus consider a function $\mathbf{u}(\mathbf{x}, t)$ that is defined on V only. Depending on the specific circumstances, the function $\mathbf{u}(\mathbf{x}, t)$ might have certain properties. For example, it might be bounded, square integrable, continuous, or differentiable with respect to the spatial variable **x**. In a mathematically rigorous description, the precise function space for $\mathbf{u}(\mathbf{x}, t)$ needs to be specified. However, this is beyond the scope of this book. Thus, in the following, we avoid full mathematical rigor, and instead require all functions to be "smooth enough" for the intended use. The space of $\mathbf{u}(\mathbf{x}, t)$ for fixed t is in general an infinite dimensional vector space, since we can find infinitely many linearly independent elements that are contained in it. This can be seen by considering the one dimensional case $d = m = 1$ with bounded region $V = [-\pi, \pi]$. Then, for a given t, the function space for $u(x, t)$ will typically at least contain the infinitely many linearly independent functions $v_j(t) \sin(jx)$ for $j \in \mathbb{Z}$, and arbitrary coefficients $v_j(t) \in \mathbb{R}$.

Using the notation $\mathbf{u}(t)$ for the distribution specified by $\mathbf{u}(\mathbf{x}, t)$ at a given t, one can express the dynamical equation of the system as

$$\dot{\mathbf{u}}(t) = F(\mathbf{u}(t)),$$
$$\mathbf{u}(0) = \mathbf{u}_0. \tag{4.58}$$

Here F is an operator that maps the spatial distribution $\mathbf{u}(t)$ to another spatial distribution, thus specifying the dynamics of the system. The initial condition $\mathbf{u}_0$ describes, for instance, the initial distribution of the substances in the chemical example mentioned above.

It is often appropriate to consider a discretized version of Equation 4.58, by assuming that $\mathbf{u}(\mathbf{x}, t)$ is well enough approximated if one specifies it at N discrete spatial positions $\mathbf{x}_j$. Often the positions $\mathbf{x}_j$ correspond to points on a regular grid. Using the notation $\mathbf{u}_j(t) = \mathbf{u}(\mathbf{x}_j, t)$, Equation 4.58 can be replaced by the system of equations

$$\dot{\mathbf{u}}_j(t) = F_j(\mathbf{u}_1(t), \ldots, \mathbf{u}_N(t)),$$
$$\mathbf{u}_j(0) = \mathbf{u}_{0j}, \tag{4.59}$$

where the operators F_j are in general different. Formally, Equation 4.59 is a system of ordinary differential equations for the Nm-dimensional state vector

$$\mathbf{U} = (\mathbf{u}_1, \ldots, \mathbf{u}_N) \in \mathbb{R}^{Nm}. \tag{4.60}$$

Note that in our terminology we strictly distinguish the term *spatially discrete* from the similar term *(time) discrete*, which we introduced in Section 1.1.1.

4.2.2 Terminology of Coupling Schemes

Depending on the specific form of the operator F in Equation 4.58 a number of different cases are commonly distinguished.

In the simplest case, F is spatially diagonal, i.e., it is of the form

$$F\left[\mathbf{u}\right]\left(\mathbf{x}\right) = \mathbf{f}\left(\mathbf{x}, \mathbf{u}\left(\mathbf{x}\right)\right), \tag{4.61}$$

and Equation 4.58 can be written as

$$\dot{\mathbf{u}}\left(\mathbf{x}, t\right) = \mathbf{f}\left(\mathbf{x}, \mathbf{u}\left(\mathbf{x}, t\right)\right). \tag{4.62}$$

In other words, in this trivial case, the full system is simply a collection of uncoupled subsystems for each point $\mathbf{x}$. The function $\mathbf{f}$ describes the local dynamics of the system.

In this case, one simple and common way to introduce a coupling scheme in Equation 4.62 is by adding a term that contains one or more spatial derivatives, such as

$$\dot{\mathbf{u}}\left(\mathbf{x}, t\right) = \mathbf{f}\left(\mathbf{x}, \mathbf{u}\left(\mathbf{x}, t\right)\right) + \mathbf{H}\left(\mathbf{x}, \left[\partial^{\alpha_1}\mathbf{u}\right]\left(\mathbf{x}\right), \ldots, \left[\partial^{\alpha_q}\mathbf{u}\right]\left(\mathbf{x}\right)\right). \tag{4.63}$$

Here we use the multiindex notation $\alpha_k = (\alpha_{k1}, \ldots, \alpha_{kd})$ with $\alpha_{kj} \in \mathbb{N}_0$ and $\partial^{\alpha_k} = \partial^{\alpha_{k1}} \cdots \partial^{\alpha_{kd}}$. Since the coupling term only depends on the knowledge of the function $\mathbf{u}$ and its derivatives at the local point $\mathbf{x}$, we speak of a *local coupling* scheme.

Equation 4.63 is a partial differential equation. Many physical systems, such as are found in continuum mechanics, fluid mechanics, or optics, are described by equations of this form. In the context of chemical substances we can, for example, consider the diffusion of individual species in space, which is described by a diffusion equation of the form

$$\partial_t \mathbf{u}\left(\mathbf{x}, t\right) = \mathbf{D}\Delta \mathbf{u}\left(\mathbf{x}, t\right). \tag{4.64}$$

Here Δ is the Laplace operator

$$\Delta = \partial_1^2 + \cdots + \partial_d^2, \tag{4.65}$$

and $\mathbf{D} = \mathrm{diag}(D_1, \ldots, D_m)$ is the $m \times m$ diffusion matrix, whose diagonal elements are the diffusion constants D_k for concentration u_k on the diagonal.

A second common way to introduce coupling in Equation 4.62 is by adding a *non-local* coupling term in the form

$$\dot{\mathbf{u}}\left(\mathbf{x}, t\right) = \mathbf{f}\left(\mathbf{x}, \mathbf{u}\left(\mathbf{x}, t\right)\right) + \int d\mathbf{x}'\, \mathbf{G}\left(\mathbf{x}, \mathbf{x}', \mathbf{u}\left(\mathbf{x}, t\right), \mathbf{u}\left(\mathbf{x}', t\right)\right). \tag{4.66}$$

This is an integro-differential equation and the function $\mathbf{G}$ is called the *coupling kernel*.

In the simplest case, $\mathbf{G}$ is independent of both $\mathbf{x}$ and $\mathbf{x}'$, thus providing a *global coupling*. In the example of the dynamics of chemical concentrations used above, global coupling can arise if the average concentration u_k of a specific chemical k over some region V enters the dynamical equations, resulting in the form

$$\dot{\mathbf{u}}\left(\mathbf{x}, t\right) = \mathbf{f}\left(\mathbf{x}, \mathbf{u}\left(\mathbf{x}, t\right)\right) + \frac{\mathbf{K}}{|V|} \int_V d\mathbf{x}' \, u_k\left(\mathbf{x}'\right), \tag{4.67}$$

where $\mathbf{K} \in \mathbb{R}^m$ is a coupling vector. Another example of a globally coupled system is the Kuramoto model discussed in Section 4.1.

When the coupling kernel assumes the form

$$\mathbf{G}\left(\mathbf{x} - \mathbf{x}', \mathbf{u}\left(\mathbf{x}, t\right), \mathbf{u}\left(\mathbf{x}', t\right)\right) \quad \text{or} \quad \mathbf{G}\left(\mathbf{x}, \left|\mathbf{x} - \mathbf{x}'\right|, \mathbf{u}\left(\mathbf{x}, t\right), \mathbf{u}\left(\mathbf{x}', t\right)\right),$$

the terms *homogeneous coupling* and *isotropic coupling* are also used, respectively.

In the spatially discrete case, both coupling schemes in Equations 4.63 and 4.66 can be written in the form

$$\dot{\mathbf{u}}_j\left(t\right) = \mathbf{f}_j\left(\mathbf{u}_j\right) + \sum_{\substack{k=1 \\ k \neq j}}^{N} \mathbf{g}_{jk}\left(\mathbf{u}_j, \mathbf{u}_k\right). \tag{4.68}$$

In this case, a coupling scheme is called *local* if the coupling functions $\mathbf{g}_{jk}$ vanish, whenever the points $\mathbf{x}_j$ and $\mathbf{x}_k$ are not nearest neighbors. Similarly, if $\mathbf{g}_{jk}$ does not explicitly depend on the indices j and k, we speak of *global coupling*.

In the special case when $\mathbf{g}_{jk}\left(\mathbf{u}_j, \mathbf{u}_k\right)$ is linear in $\mathbf{u}_j - \mathbf{u}_k$, the term *diffusive coupling* is often used, which is in contrast to the term *direct coupling*, which denotes the situation of $\mathbf{g}_{jk}\left(\mathbf{u}_j, \mathbf{u}_k\right)$ being linear in $\mathbf{u}_k$ and independent of $\mathbf{u}_j$ (Aronson et al. 1990).

4.3 Chimera States

The topic of *chimera states* has fascinated researchers in nonlinear dynamics since their theoretical discovery by Kuramoto and Battogtokh (2002), and has led to a substantial reconsideration of previously implicit assumptions about the nature of synchronization in high-dimensional systems.

While at first chimera states had been considered just a theoretical curiosity without much consequence for real-world systems, the recent evidence for experimental realizations (Tinsley et al. 2012; Hagerstrom et al. 2012) has further boosted the research activity in this area, and has pushed the subject to the level of interdisciplinary prominence.

At the time of writing this book, the subject has not yet settled down, and even a generally accepted definition of what a chimera state actually is has not been reached within the communities involved. The purpose of this section is to give a brief overview of the established results in the field, and to present some of the more recent developments. For a recent review on the subject we refer the reader to Panaggio and Abrams (2015).

In the following, we first describe the numerical phenomenology of the "classical" chimera state, before considering more recent experimental evidence, and concluding with some analytical considerations.

4.3.1 Numerical Phenomenology of the Classical Chimera State

The Classical Chimera State in One Dimension

Up to the early 2000s it was generally believed that a system of identical coupled oscillators should either be in a coherent phase-locked state, where all oscillators follow the same dynamics up to some phase shift, or in an incoherent state where the phases of any individual oscillators move in quasiperiodic fashion.

However, in 2002, Kuramoto and Battogtokh showed that even in a simple system of identically coupled oscillators on a ring, coherent and incoherent states can coexist in well-separated spatial regions.

A classical example of such a fascinating state is shown in Figure 4.10. It depicts a snapshot of phases for a collection of phase oscillators along the x axis, and before we even introduce the details of the model, we can appreciate the visually clear separation between a domain in the centre, where the phase changes discontinuously with x, and a domain at the boundary with a continuous phase variation. Note that since periodic boundary conditions are employed, the parts on the left and right boundaries form a single connected continuous region.

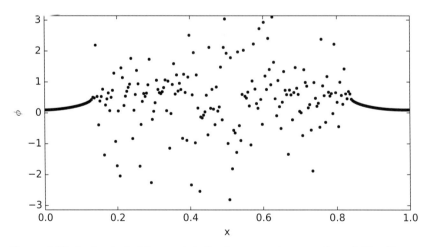

Figure 4.10 A classical chimera state for the governing equations 4.69 with $\kappa = 4.0$ and $\alpha = 1.457$ (Kuramoto and Battogtokh, 2002). The Python computer code that produces this graph is shown in Figure 4.12.

Figure 4.10 is only a snapshot at one instant in time, but if we looked at a movie of this state, we would observe that all the points in the continuous region coherently move with the same constant phase velocity. This is in contrast to the points in the discontinuous central region. There, even neighboring points have different instantaneous phases and different average velocities. Each point in this region follows its individual phase-slip dynamics similar to a driven but unlocked phase oscillator. This gives rise to the unlocked and thus incoherent dynamics in the central region. This separation into coherent and incoherent parts is numerically stable even for extremely long integration times.

Abrams and Strogatz (2004) coined the term "chimera states" for this coexistence of two seemingly incompatible states in a single system. This is in reference to the *chimera* creature of Greek mythology, which has two or more heads from "incompatible" animals (typically goat, lion and snake) attached to a single body.

In order to understand how chimera states arise, let us study the model that underlies the dynamics of Figure 4.10 in more detail. The governing equation for the function $\varphi(x, t)$ is given by (Kuramoto and Battogtokh, 2002; Abrams and Strogatz, 2006)

$$\frac{\partial}{\partial t}\varphi(x, t) = \omega - \int_0^1 G(x - x') \sin(\varphi(x, t) - \varphi(x', t) + \alpha) \, dx' . \quad (4.69)$$

Here, space is assumed to be a one-dimensional ring with periodicity one, so we choose $x \in [0, 1)$ with periodic boundary conditions. The dynamical phase variable $\varphi(x, t) \in [-\pi, \pi)$ has the usual 2π periodicity.

The angular frequency parameter ω has no influence on the dynamics of the system, and we have the option to eliminate it by transforming to comoving coordinates $\hat{\varphi} = \varphi - \omega t$. The integral term in Equation 4.69 is the source of coupling between different space points. It contains a kernel function $G(y)$ that was chosen to be proportional to $e^{-\kappa|y|}$ by Kuramoto and Battogtokh (2002). With the normalization $\int_0^1 G(y) \, dy = 1$ the kernel is given by

$$G(y) = \frac{\kappa}{2\left(1 - e^{-\frac{\kappa}{2}}\right)} e^{-\kappa|y|} . \quad (4.70)$$

This kernel describes a nonlocal coupling between different positions x, which is in marked contrast to the previously discussed cases of local coupling and global coupling. The parameter κ quantifies how quickly the coupling decays with increasing distance between oscillators, with the two limit cases being (i) $\lim_{\kappa \to 0} G(y) = 1$ (globally coupled case) and (ii) $\lim_{\kappa \to \infty} G(y) = \delta(y)$ (uncoupled case).

Apart from κ, the only other free parameter in the model equation 4.69 is the *phase lag* α. This parameter was originally introduced by Sakaguchi and Kuramoto (1986) for the globally coupled Sakaguchi–Kuramoto model, which one obtains as a special case of Equation 4.69 if one sets $G(y) = 1$. For Figure 4.10 the parameters $\kappa = 4$ and $\alpha = 1.457 < \pi/2$ were used.

To solve Equation 4.69, one needs to supply an initial condition $\varphi(x, 0) = \varphi_0(x)$. For simple initial conditions, for example $\varphi_0(x) = 0$, one obtains the trivial phase-locked solution $\varphi_L(x, t) = [\omega - \sin(\alpha)]t$. It can be shown that this solution is linearly stable for $|\alpha| < \pi/2$. In order to get the desired chimera states, it is thus crucial to choose appropriate initial conditions.

At this point, it is important to note that the function $\varphi(x, t)$ is not required to be smooth or even continuous in the space variable x for the model 4.69 to be well defined. We use this freedom in our choice to impose a discontinuous $\varphi_0(x)$. A good choice is to select the initial phase randomly at each point x, for example according to the rule (Abrams and Strogatz, 2006)

$$\varphi_0(x) = 2\pi e^{-30\left(x - \frac{1}{2}\right)^2} r(x), \qquad (4.71)$$

where $r(x)$ is a uniformly distributed random variable on $[-1/2, 1/2)$. The effect of this "single-hump" initial condition is that the initial phase is essentially fixed to zero at the boundary, while close to the centre point $x = 1/2$ the phases are almost arbitrarily distributed. This is shown in Figure 4.11.

Numerical Implementation

A good understanding of chimera states can be obtained by "playing" with them numerically on a computer. As a starting point, in Figure 4.12 we provide a short Python program that produces Figure 4.10. We stress that the primary purpose of this code is educational, and in the interest of transparency we refrained from optimizing it for speed and accuracy.

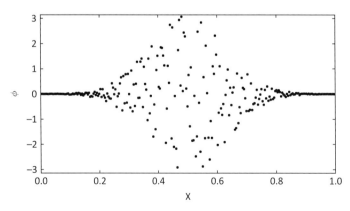

Figure 4.11 Initial conditions for the appearance of the chimera state, according to Equation 4.71.

```
1   import matplotlib.pyplot as plt
2   import numpy as np
3
4   alpha = 1.457
5   kappa = 4.0
6
7   N = 256
8   dt = 0.025
9   Tsteps = 8000
10
11  X = np.linspace(0, 1, N, endpoint=False)
12  phi = (np.random.random(N)-0.5) * 2*np.pi * np.exp( -30*(X-0.5)**2 )
13  phidot = np.zeros(N)
14
15  G0 = kappa / (2 * (1-np.exp(-kappa/2)) ) * np.exp(-kappa * X[:N//2+1])
16  G = np.concatenate((G0[:-1],G0[:0:-1],G0[:-1],G0[:0:-1]))
17
18  for j in range(Tsteps):
19      for i in range(N):
20          phidot[i] = -np.sum(np.sin(phi[i]-phi+alpha) * G[N-i:2*N-i])/N
21      phi += dt * phidot
22
23  plt.figure(figsize=(8,4))
24  plt.plot(X, (phi + np.pi) % (2*np.pi)-np.pi, "k.")
25  plt.ylim(-np.pi, np.pi)
26  plt.xlabel("x")
27  plt.ylabel("$\varphi$")
28  plt.savefig("classic_chimera.pdf", bbox_inches="tight")
```

Figure 4.12 Python code used to produce the classical chimera states of Figure 4.10.

The code implements the mathematical model 4.69 by discretizing the space-dependent phase variable $\varphi(x, t)$ into $N = 256$ individual coupled oscillators $\varphi_k(t)$ located at positions $x_k = k/N$ for $k = 0, \ldots, N - 1$. While this step seems obvious, it does raise some slight concerns, since the function $\varphi(x, t)$ is in general discontinuous and it is thus not clear in what sense a single oscillator $\varphi_k(t)$ can approximate the phase function $\varphi(x, t)$ over some space interval of length $1/N$. One solution that has been suggested (see footnote 6 in Panaggio and Abrams 2015) is to consider the space-continuous model equation 4.69 to be merely an "abbreviation" of the discretized model in the limit of large N. The discretized model is explicitly given by

$$\dot{\varphi}_j = \omega - \frac{1}{N} \sum_{k=0}^{N-1} G\left(x_j - x_k\right) \sin\left(\varphi_j - \varphi_k + \alpha\right). \tag{4.72}$$

Under discretization, the single-hump initial condition 4.71 becomes a random initialization of the individual oscillators (see line 12 of the code in Figure 4.12 and the resulting graph in Figure 4.11). The integration kernel G of Equation 4.70 is constructed in lines 15–16, and then the sum of Equation 4.72 is implemented in line 20. The simple forward Euler integrator in line 21 finally produces the dynamical evolution of the state vector φ and yields, after a sufficient number of time steps, the classical chimera state as reported in Figure 4.10.

While our little program in Figure 4.12 produces the classical chimera state in a transparent way, it leaves out a number of obvious opportunities for optimizations, aside from the obvious drawbacks of using a language, such as Python, that is highly inadequate for proper numerical simulations or analyses of large systems. The main problem of the code is that the number of calculations per time step scales like N^2, since the summation in line 20 requires $\mathcal{O}(N)$ operations, and the line itself is iterated N times through the loop defined in line 19. With increasing N, the program is therefore not very efficient. A key observation that helps writing faster code is that the sum in Equation 4.72 can be rewritten as

$$\sum_{k=1}^{N} G\left(x_j - x_k\right) \sin\left(\varphi_j - \varphi_k + \alpha\right) = \Im\left(e^{i\varphi_j} e^{i\alpha} \sum_{k=0}^{N-1} e^{-i\varphi_k} G\left(\frac{j-k}{N}\right)\right). \quad (4.73)$$

Then the sum on the right-hand side is simply a convolution of the two vectors $\left(e^{-i\varphi_0}, \ldots, e^{i\varphi_{N-1}}\right)$ and $(G(0), G(1/N), \ldots, G((N-1)/N))$. If N is conveniently chosen to be a power of 2, then the Fast Fourier Transform (FFT) method can be employed to calculate this convolution very efficiently (Press et al. 2007). This works by first calculating the FFTs of the discretized versions of both vectors, then forming the product of the two Fourier transforms, and finally applying a backward FFT on the result of this product. Using this method the number of operations per time step scales with the complexity of the FFT, which is $N \log N$. For sufficiently large N this is much faster than the previous complexity of N^2. Another obvious optimization of our code is to use a more accurate integration algorithm than the primitive forward Euler method, such as a Runge–Kutta integration scheme.

Choice of the Kernel

We have demonstrated that the simple model in Equation 4.69 leads numerically to chimera states when using the kernel $G(y)$ of Equation 4.70. This could generate concerns that the chimera states critically depend on the choice of the kernel. Fortunately, this is not the case, and other kernels are possible, as long as they give rise to nonlocal coupling. Specifically, the functions (Abrams and Strogatz 2004)

$$G(y) = \frac{1}{2\pi}(1 + A \cos x) \quad (4.74)$$

for $0 \leqslant A \leqslant 1$, or the normalized *top hat* function

$$G(y) = \begin{cases} \frac{1}{2R} & \text{for } |y| < R \\ 0 & \text{otherwise} \end{cases} \quad (4.75)$$

with $0 \leqslant R \leqslant 1/2$ give numerically very similar results to those reported in Figure 4.10.

Collapse of Classical Chimera for Finite N

The classical chimera seeks to model an idealized situation where the number of oscillators goes to infinity. However, it is clear that numerically the limit $N \to \infty$ can only be approximated. It is therefore a valid question, what finite-size effects are expected from numerical simulations.

This problem was investigated by Wolfrum and Omel'chenko (2011), and it was found that for small N ($N < 45$) the chimera state after some time collapses and disappears. What is interesting is that, until just before the collapse, the chimera appears perfectly "healthy," but then suddenly flatlines within a short time. After the collapse, the dynamics proceed in a fully synchronized state. The average time until this collapse happens is numerically found to scale exponentially with N. This scaling and the suddenness of the collapse are analogous to the type-II transient turbulence state in nonlocally coupled map lattices (Crutchfield and Kaneko 1988). The exponential scaling of the collapse time means that the collapse will typically not be observed for reasonably large N. However, it also suggests that for finite N, all classical chimera states are transients.

4.3.2 Classical versus Generalized Chimera States

The states described in the previous subsection constitute are the so-called *classical* chimera states. To distinguish them from the many variations that have been proposed and investigated, we list some of their established properties:

(i) The state subdivides the phase space into a coherent and an incoherent part, which coexist.
(ii) The phase space is composed of a collection of identical oscillators.
(iii) The initial conditions leading to the state have a measure larger than zero.
(iv) The dynamical system is continuous in time.
(v) The coupling is homogeneous, i.e., it does not subdivide the oscillators into sub-communities.
(vi) The oscillators are spatially arranged along a (1-dimensional) ring.
(vii) The oscillators are phase oscillators.
(viii) The limit of $N \to \infty$ is considered.
(ix) Chimera states are stable in space and time for large N.
(x) The number of oscillators in both the coherent and incoherent parts scales linearly in N in the limit of $N \to \infty$.
(xi) There are no time-delay terms in the system.
(xii) The system has nonlocal coupling.

Since their initial discovery, a multitude of other systems with similar features have been studied and analyzed, which give rise to patterns that differ from the classical chimera in at least one (but usually many) of the above-listed properties. Therefore, a question arises: which features of the classical chimera state should be taken as the defining properties for the term "chimera" in a more general context?

Currently, only the first two seem to be generally accepted. Property (i) is considered indispensable because the peculiarity of the coexistence of coherent and incoherent parts was the original reason for the name of these states. However, the precise definitions of the terms "coherent" and "incoherent," and even of the term "parts" are not at all completely obvious and actually require some clarification.

Requirement (ii) is highly desirable on theoretical grounds, since we want to be able to distinguish the chimera state from the well-understood state of a phase-synchronized cluster of nonidentical oscillators, which we also encountered in the transition to synchronization in the standard Kuramoto model in Section 4.1. However, true identity of oscillators is almost impossible to achieve in any experimental setup. In experiments, the requirement is therefore often relaxed to the condition that unavoidable differences between subunits should have no qualitative influence on the observed dynamics.

Requirement (iii) is often implicitly acknowledged. The chimera states are typically not the only attracting states, and might for example compete with the completely locked state. Their basins of attraction might therefore be small, but it is reasonable to require that not only exceptional initial conditions give rise to chimera states, since otherwise they would not be robust and their relevance could be difficult to justify.

All other conditions in the list have been modified in the past to demonstrate new types of chimera states. In the following, we adapt a liberal terminology, which accepts the term "chimera" in a fairly broad sense, as long as the requirements (i) and (ii) are fulfilled, and we give a short overview of some of the many variations of chimera states that have been suggested.

Chimeras in Two and More Space Dimensions

As observed by Shima and Kuramoto (2004), a straightforward generalization of the model Equation 4.69 in two spatial dimensions on an infinite plane gives rise to a coherent spiral pattern, with an incoherent circular region at the center of the spiral, as shown in Figure 4.13. This effect is analogous to the classical chimera in one dimension, but it is interesting to note that it does not require the use of periodic boundary conditions. The synchronous outer region can be arbitrarily large.

Chimeras in Globally Coupled Chaotic Maps

As we wrote above, all other properties are not generally considered necessary to define a chimera state. For instance, the requirement of a time-continuous system is

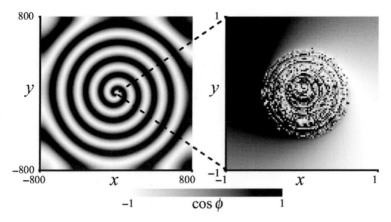

Figure 4.13 Coherent spiral structure with circular chimera at the center. Reprinted with permission from Shima and Kuramoto (2004).

usually not taken as essential, and iterated maps might be considered as well. It was pointed out by Yeldesbay et al. (2014) that a state "reminiscent of a chimera" in maps was discovered already in Kaneko (1990) long before the work by Kuramoto and Battogtokh on the classical chimera state. Kaneko studied a simple system of N globally coupled identical chaotic maps (Kaneko 1990, 2015) given by

$$x_{n+1}(i) = (1 - \varepsilon) f(x_n(i)) + \frac{\varepsilon}{N} \sum_{j=1}^{N} f(x_n(j)), \qquad (4.76)$$

where i is the index of the element and $f(x) = 1 - ax^2$ is the logistic map. The only parameters in the system are therefore the nonlinearity a, the coupling strength ε and the sample size N.

Kaneko found that, for certain parameters and initial conditions, about half of the elements are perfectly synchronized within one big cluster, while all the other elements evolve incoherently with respect to the big cluster and also do not mutually synchronize among themselves. A plot of this state using Kaneko's parameters is shown in Figure 4.14. Historically, we may consider this as the first demonstration of a generalized chimera state, although the connection to chimera states was only made more than twenty years later. In hindsight, Kaneko's chimera shows that a simple global coupling of identical elements is sufficient to fulfil the first three criteria in our list, if we allow for iterated maps instead of time-continuous systems.

Chimera States in Hierarchical Populations

Abrams et al. (2008) presented a simple model where the initial population is divided into two subpopulations. The essential feature in this case is that the coupling of each element to members of its own population is stronger than the

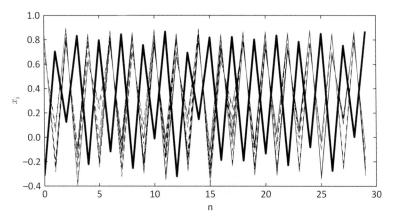

Figure 4.14 Chimera state in chaotic map lattices, using Equation 4.76 with parameters $a = 1.95$, $\varepsilon = 0.2$, and $N = 50$. There are 24 synchronized elements following the thick line, and 26 incoherent elements following the thin, dashed lines. The parameters correspond to the ones used by Kaneko in Figure 2(c) of (Kaneko 1990).

coupling to members of the other populations. This is, for example, motivated by the two hemispheres of the brain. Due to the groove between the hemispheres, neurons in the left half are more strongly connected with other neurons in the same hemisphere and less strongly connected with neurons in the right half.

An interesting phenomenon in this context is what is called unihemispheric slow-wave sleep, i.e., the ability of some animals to sleep with one half of the brain while the other half remains alert. This is in contrast to normal sleep, where both halves of the brain show reduced consciousness.

In a simplified picture, a sleeping brain can be associated with a largely synchronized neuron population, while in a brain that is awake the neurons are not synchronized and fire chaotically. It can therefore be argued that unihemispheric sleep can be associated with a generalized chimera state, where one subpopulation of oscillators is synchronized, while the other is not.

Such a basic idea can be captured in the following simple model

$$\dot{\varphi}_j^\sigma = \omega - \sum_{\sigma'=1}^{2} \frac{K_{\sigma\sigma'}}{N_{\sigma'}} \sum_{k=1}^{N_{\sigma'}} \sin\left(\varphi_j^\sigma - \varphi_k^{\sigma'} + \alpha\right), \qquad (4.77)$$

where φ_j^σ is the phase of the jth oscillator in population σ of size N_σ, with $\sigma \in \{1, 2\}$. We assume equal population sizes $N_1 = N_2 \equiv N$. The coupling matrix $K_{\sigma\sigma'}$ is such that the conditions

$$K_{in} \equiv K_{11} = K_{22} > K_{out} \equiv K_{12} = K_{21} > 0 \qquad (4.78)$$

hold. The other parameters ω and α are as in the classical chimera model of Equation 4.69.

In this context, it is useful to introduce the definition of *indistinguishable oscillators* following Ashwin and Burylko (2015), which captures the idea that each oscillator has the same levels of input link strengths, and for each level the same number of connections. In the case at hand, each oscillator has $N - 1$ input links of magnitude K_{in} and N links of magnitude K_{out}. In a slight generalization of this term, we can also define a system to consist of *indistinguishable subunits* if it is invariant under a permutation symmetry group $G \subset S_N$ such that for any pair of subunits (j, k) there exists a $g \in G$ with $g(j) = k$ (i.e., G acts transitively on the set of subunits). In the model of Equation 4.77 the group G contains all permutations that either fully swap the two subpopulations or only permute the subpopulations among themselves. The classical chimera also shows this property of indistinguishable oscillators, with the group G being the symmetry group of the ring (rotations and reflections).

When the model of Equation 4.77 is numerically simulated, it does indeed show the desired simultaneous effects of complete synchronization in one subpopulation, and incoherence in the other. In contrast to the classical chimera model, the subpopulations as such are already predefined through the hierarchical topology of the connections and the coupling is not homogeneous. Feature (v) of our list is therefore not fulfilled. The advantage of this model is that it can be solved exactly, and that its solution can be generalized to an arbitrary number of communities of arbitrary sizes (Pikovsky and Rosenblum 2008).

Chimera States in Globally Coupled Systems with Time Delay

Since chimera states are observed in globally coupled chaotic maps and in phase oscillators with hierarchical coupling, the question arises as to whether they can also be observed in a globally and homogeneously coupled ensemble of identical oscillators. Surprisingly, it was found that if a *time-delay* term is included in the definition of the individual oscillator dynamics, then a chimera state is indeed possible. This was demonstrated by Yeldesbay et al. (2014) using the model equations

$$\dot{\varphi}_k = \omega + \beta \sin \left(\varphi_{\tau,k} - \varphi_k \right) + \frac{\varepsilon}{N} \sum_{j=1}^{N} \sin \left(\varphi_j - \varphi_k + \alpha \right), \qquad (4.79)$$

where ω, α, ε, and τ are real parameters, and the notation $\varphi_{\tau,k}(t) = \varphi_k(t - \tau)$ for the time-delayed term has been used. In this model, all oscillators are fully equivalent and therefore the system is invariant under all possible permutations of oscillators. However, for certain combinations of parameters the system remarkably splits into two subsets of oscillators. In one subset all oscillators are fully synchronized, while in the other all oscillators are unsynchronized.

As in the classical chimera state, a fully synchronized solution of the form

$$\varphi_1 = \ldots = \varphi_N = \Omega t \tag{4.80}$$

exists. The parameter Ω is obtained by inserting Equation 4.80 into Equation 4.79, obtaining

$$\Omega = \omega - \beta \sin(\Omega \tau) + \varepsilon \sin(\alpha). \tag{4.81}$$

However, the synchronized solution Equation 4.80 is not necessarily stable, and it is instructive to understand why this is the case. If one perturbs the first two oscillators, so that $\varphi_{1,2}(t) = \Omega t \pm \delta(t)$ with some small perturbation $\delta(t)$, then the equation for the evolution of δ is, to first order,

$$\dot{\delta}(t) = \beta \cos(\Omega \tau)(\delta(t - \tau) - \delta(t)) - \varepsilon \cos(\alpha)\delta(t). \tag{4.82}$$

With the Ansatz $\delta(t) = \delta_0 e^{\lambda t}$, this translates into the requirement that the parameter λ fulfils the equation

$$F(\lambda) \equiv \beta \cos(\Omega \tau)\left(e^{-\lambda \tau} - 1\right) - \varepsilon \cos(\alpha) - \lambda = 0. \tag{4.83}$$

From Equation 4.83, one concludes that $F(\lambda)$ is continuous in λ and that $F(0) = -\varepsilon \cos(\alpha)$. On the other hand, one also observes that $\lim_{\lambda \to \infty} F(\lambda) = -\infty$. As a consequence, if $F(0) > 0$, there will be at least one value $\lambda_* > 0$ such that $F(\lambda_*) = 0$. This leads to the existence of an exponentially growing perturbation $\delta(t) = \delta_0 e^{\lambda_* t}$. Taking everything together, one concludes that the solution of Eq. 4.80 is indeed unstable if $\varepsilon \cos(\alpha) < 0$.

Instability of the fully synchronized solution does not necessarily lead to a chimera-like state, and other states – for example, a fully unsynchronized state, or a state with multiple synchronization clusters – could also arise. Numerical evidence, however, shows that, at least for certain parameter values, the ensemble of oscillators splits up into a single synchronized cluster and an unsynchronized "cloud" (Yeldesbay et al. 2014).

Weak Chimera States

So far, we have concentrated on a number of examples of coexistence of coherent and incoherent parts (clusters or domains) in a dynamical system, which are associated with the loose term *chimera* in the literature, but we have avoided giving a precise definition. In this subsection we define the term *weak chimera*, which has been suggested by Ashwin and Burylko (2015) for a class of coupled oscillatory systems, but that deliberately does not claim to cover chimeras in the broader sense. Such a restrictive, and mathematically precise, definition has the advantage that it is possible to give exact conditions under which weak chimeras do (or do not) occur.

Following Ashwin and Burylko (2015), we consider a system of N identical coupled phase oscillators, i.e., the phase space of the system is given by a torus $\mathbb{T}^N = [-\pi, \pi)^N$ and the evolution equations for a trajectory $(\varphi_1(t), \ldots, \varphi_N(t))$ on $\mathbb{T}^N$ are of the form

$$\dot{\varphi}_j = \omega + \sum_{k=1}^{N} K_{jk} g\left(\varphi_j - \varphi_k\right). \tag{4.84}$$

Here ω is the common frequency, K_{jk} is the coupling strength, and $g(\varphi)$ is a 2π-periodic smooth coupling function.

The trajectory phase functions $\varphi_j(t)$ are only defined modulo 2π, and in the following we will always use continuous representatives. For a given trajectory of Equation 4.84, the oscillator pair (j, k) is said to be *phase-synchronized* if

$$\lim_{t \to \infty} \frac{1}{t} \left(\varphi_j(t) - \varphi_k(t)\right) = 0. \tag{4.85}$$

As a further preparation, we need to define the term *chain-recurrent set*, which is slightly abstract (for a rigorous definition see Definition 1.3 of Mischaikow et al. 1995). In less formal words, a chain-recurrent set A is a flow-invariant subset of the phase space of a dynamical system, such that for any point $x \in A$ and for any $\varepsilon, T > 0$ there exists an (ε, T)-chain from x to x in A. In this context, an (ε, T)-chain from x to y in A is a finite sequence $(x_1 = x, x_2, \ldots, x_n, x_{n+1} = y)$ of points of A such that a trajectory which starts out at point x_k at time 0 will visit an ε-neighborhood of point x_{k+1} at a time $t_k > T$. Connected chain-recurrent sets arise in practice often as (abstract) ω-limit set of some initial point x (Franke and Selgrade 1976). With these premises we are in a position to define a *weak chimera* as follows.

Definition (Ashwin and Burylko 2015): A *weak chimera* of a system of N coupled phase oscillators is a flow-invariant, connected and chain-recurrent set $A \subset \mathbb{T}^N$ such that for any orbit in A there exist indices j, k, l with the properties that the oscillator pair (j, k) is phase-synchronized and the oscillator pair (j, l) is not phase-synchronized.

It should be stressed that this definition weakens the previous notion of property (i), i.e., the separation of the system into incoherent and coherent subpopulations of oscillators as a requirement for chimera states. In fact, oscillator l in the definition could well be part of a cluster of oscillators with which its phase synchronizes. Furthermore, the definition only applies to coupled phase oscillators with finite N, and in particular it does not imply that the set A is stable.

However, using this definition it is now possible to study the existence and stability of weak chimeras in a number of coupled oscillator networks. Using a coupling function of the form

$$g(\varphi) = -\sin(\varphi + \alpha) + r\sin(2\varphi) \tag{4.86}$$

with the usual parameter α and a second Fourier component proportional to r, Ashwin and Burylko (2015) showed that stable weak chimeras exist for a number of small networks of indistinguishable oscillators.

4.3.3 Experimental Implementation of Chimera States

Chimera states were first studied theoretically as intriguing objects, but their experimental significance was doubtful for a long time, especially given the fact that they intrinsically require a system made of identical subunits.

As a matter of fact, experimental evidence failed to materialize for over a decade, and the perceived challenges to the experimental observation of classical chimeras seemed difficult to overcome. In fact, first one is required to implement a large ensemble of almost identical oscillators. Second, the requirement on non-global coupling can be challenging in an experimental context.

Indeed, the classical chimera state described by Equation 4.72 has still not been fully reproduced experimentally. However, many of the generalized chimeras discussed in Section 4.3.2 are more amenable to experimental implementation and, since the first such success in 2012, experimental activities have increased dramatically. In the following, we give a brief overview of some interesting experimental setups, without any claim for completeness.

Chimeras in Coupled Chemical Oscillators

The first experimental observation of a chimera state was obtained by Tinsley et al. (2012) in a system of discrete chemical oscillators on the basis of the Belousov–Zhabotinsky reaction (Winfree 1984; Pikovsky 1981). A population of 40 individual oscillators was divided into two subpopulations, with the current intensity $I_i^\sigma(t)$ of each oscillator measured via a CCD camera. Here, $\sigma = \{1, 2\}$ indicates the subpopulation and i is the index within that subpopulation. A chemical oscillator can be influenced by changing the feedback light intensity P_i^σ irradiated on it. Every three seconds, the feedback light is calculated by a computer using the formula

$$P_i^\sigma = P_0 + k_\sigma \left(\bar{I}^\sigma(t - \tau) - I_i^\sigma(t)\right) + k_{\sigma\sigma'} \left(\bar{I}^{\sigma'}(t - \tau) - I_i(t)\right). \tag{4.87}$$

Here, k_σ is the intrapopulation coupling, $k_{\sigma\sigma'}$ is the interpopulation coupling, and $\bar{I}^\sigma(t - \tau)$ is the average intensity of subpopulation σ at a time τ before the

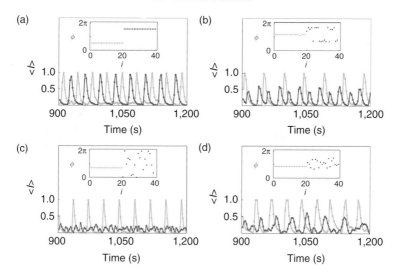

Figure 4.15 Experimental time traces of the average light intensities of chemical oscillator subpopulations (Tinsley et al. 2012). By varying the inter- and intra-population couplings, a transition between the 1–1 antiphase state in panel (a), the 2–1 synchronized state in panel (b) and the chimera state in panel (c) is achieved. The dynamics in panel (d) are interpreted as a *semisynchronized* state, and could correspond to a phase-synchronized state as introduced in Equation 4.85. Reprinted with permission from Tinsley et al. (2012).

current time t. In the experiments, $\tau = 30$ s was used (Tinsley et al. 2012). As described in Section 4.3.2, the intrapopulation coupling is taken to be stronger than the interpopulation coupling. The calculated feedback is then applied to each individual oscillator via a spatial light modulator. The resulting average intensity time traces are shown in Figure 4.15. The system can exhibit 1:1 synchronization, 2:1 synchronization and, most importantly, a chimera state as demonstrated in Figure 4.15(c).

Chimeras in a Liquid-Crystal Spatial Light Modulator

The first experimental observation of a chimera state within a ring topology was done by Hagerstrom et al. (2012), using a liquid crystal spatial light modulator (SLM) consisting of $N = 256$ individual pixels.

The experimental setup is shown in Figure 4.16. Light from a LED is sent onto the SLM, and the reflected light is analyzed using a camera. This way, the current state of the SLM can be obtained, and fed to a computer. The computer then calculates the feedback strength based on the desired connection topology and passes it back onto the SLM. The update of the feedback is done at discrete times and therefore this system behaves like a coupled map lattice. Since the connection

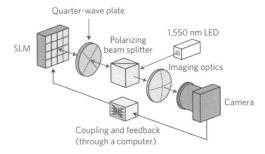

Figure 4.16 Experimental setup for demonstrating chimera states in coupled map lattices. Reprinted with permission from Hagerstrom et al. (2012).

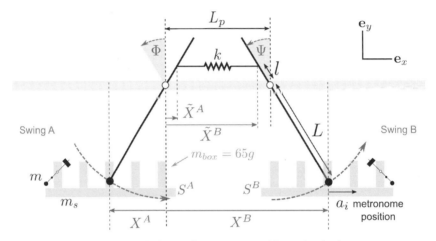

Figure 4.17 Two subpopulations of $N_1 = N_2 = 15$ mechanical metronomes are placed on two swings. The swings are mutually coupled with a spring. Reprinted with permission from Martens et al. (2013).

topology is created via a computer, arbitrary topologies are possible. Using this setup, coexistence of coherent and incoherent behavior was indeed found.

Chimera States in Mechanical Oscillator Networks

The two experiments mentioned so far used a computer at a decisive step to calculate the feedback strength. In a beautiful experiment, Martens et al. (2013) demonstrated that a computer device is not required, and chimera states can be created by fully mechanical setups.

Martens and collaborators placed 30 metronomes onto two swings, as shown in Figure 4.17. The tiny motion of each swing is sufficient to induce synchronization, if both swings are uncoupled. This effect is strikingly similar to the famous classical clock experiment by Huygens in 1665 (Bennet et al. 2002), where the tiny perturbations of a beam, on which two clocks are suspended, was able to synchronize them. In addition, Martens et al. tuned the coupling between the subcommunities on both swings by changing the elastic constant of the spring connecting the two swings. The effect of this setup is a hierarchical coupling as described in Section 4.3.2. For appropriate parameters, the dynamical system reaches a state in which all metronomes on one swing mutually synchronize, while the oscillators on the other swing oscillate incoherently.

Virtual Chimera in Time-Delayed Systems

The classical chimera exists on a continuous spatial domain, and it is therefore inherently difficult to directly implement it within an experiment. Instead, Larger et al. (2013) took a different approach and created "virtual" chimera states in a highly interesting way. They used an electronic bandpass frequency modulation delay oscillator, which (in rescaled units) implements the system of equations

$$\varepsilon \dot{x} = -\delta y - x + f\left(x\left(t-1\right)\right)$$
$$\dot{y} = x \,. \tag{4.88}$$

Because of the appearance of the term $x\left(t-1\right)$, this is a delay differential equation, and time has been rescaled to units of the delay time (which was 2.5 ms in the experiment). The function $f\left(x\right)$ is a nonlinear function whose magnitude can also be rescaled in the experiment. The parameters ε and δ are small and their values are chosen in a way to ensure that, when $f = 0$, the system represents a strongly damped linear oscillator whose characteristic decay time ε is much smaller than unity.

The experiment, as well as the numerical integration of the system, yields a time series $x\left(t\right)$. To represent this time series in a two-dimensional fashion, it is stroboscopically split into small segments of length $(1 + \gamma)$, where γ is a small parameter, and then the individual segments are stacked on top of each other. In other words, one rewrites the time variable in the form

$$t = (1 + \gamma)(n + \sigma)\,, \tag{4.89}$$

where n is an integer that grows uniformly in time, and σ is a number between 0 and 1 that represents a "virtual" space axis. The value of γ is chosen to ensure that the patterns in the resulting $\sigma - n$ plots have minimal drift. The two-dimensional plots show a striking similarity with classical chimera states, including single-headed, and multiheaded chimeras. A sample of the possible dynamics is shown in Figure 4.18.

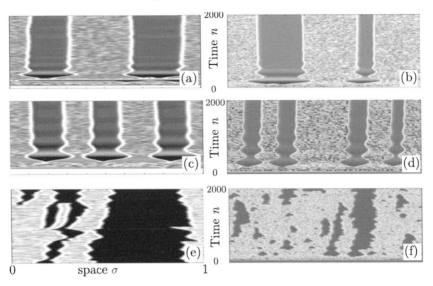

Figure 4.18 Experimentally observed frequency dynamics in an electronic oscil-lator device (left column) and in numerical simulation of system 4.88 (right column). Each point in the (σ, n) plane corresponds to one point in time accord-ing to Equation 4.89. Two-headed virtual chimera states are shown in the top row. The middle row shows experimentally observed three-headed chimeras (c) and numerically observed four-headed chimeras (d). The bottom row shows turbulent behavior. Reprinted with permission from Larger et al. (2013).

4.3.4 Theory of Chimeras

After having described the numerical and experimental phenomenology of chimera states, we now discuss how the appearance of such states can be understood on an analytical basis.

We start by first rewriting Equation 4.69 as

$$\frac{\partial}{\partial t}\varphi(x, t) = \omega - \Im\left(e^{i(\varphi(x,t)+\alpha)} \int_0^1 G(x - x') e^{-i\varphi(x',t)}dx'\right). \qquad (4.90)$$

This expression suggests the introduction of two real order parameters $R(x, t)$ and $\Theta(x, t)$, given by

$$R(x, t) e^{i\Theta(x,t)} = \int_0^1 G(x - x') e^{i\varphi(x',t)}dx', \qquad (4.91)$$

which allow one to rewrite Equation 4.90 in the form

$$\frac{\partial}{\partial t}\varphi(x, t) = \omega - R(x, t)\Im\left(e^{i(\varphi(x,t)-\Theta(x,t)+\alpha)}\right)$$
$$= \omega - R(x, t)\sin(\varphi(x, t) - \Theta(x, t) + \alpha). \qquad (4.92)$$

So far, all transformations made are exact.

We note that a change in ω has a rather trivial effect on the resulting dynamics. In fact, the change $\omega \rightarrow \omega'$ yields a corresponding solution transformation

$$\varphi (x, t) \rightarrow \varphi' (x, t) = \varphi (x, t) + (\omega' - \omega) t. \tag{4.93}$$

This means that we are effectively free to choose ω as is most convenient.

The critical step is now to look for self-consistent solutions for $R (x)$ and $\Theta (x)$ that only depend on space, but not on time (Kuramoto and Battogtokh 2002; Abrams and Strogatz 2004, 2006). Of course, time-independent solutions will in general only exist for an appropriately chosen value of ω. If such functions $R (x)$ and $\Theta (x)$ can be found, then Equation 4.92 decouples, and can be solved for each position x individually.

There are then two cases to consider. The first is $R (x)^2 \geqslant \omega^2$, for which one has a stable fixed point $\varphi^* (x)$ that fulfils the equation

$$\omega = R (x) \sin (\varphi^* (x) - \Theta (x) + \alpha) . \tag{4.94}$$

Conversely, if $R (x)^2 < \omega^2$, one obtains the typical phase-slip dynamics with periodicity

$$T (x) = \frac{2\pi}{\sqrt{\omega^2 - R (x)^2}}. \tag{4.95}$$

In this case, one can think of $\varphi (x)$ as if it were smeared over the full phase circle, as a random variable. The density function for $\varphi (x)$ is given by

$$p_x (\varphi) = (T (x) |\dot{\varphi}|)^{-1} = \frac{\sqrt{\omega^2 - R (x)^2}}{2\pi |\omega - R (x) \sin (\varphi - \Theta (x) + \alpha)|}, \tag{4.96}$$

which is analogous to the corresponding density $\hat{\rho}_t (\varphi, \omega)$ which we used in the discussion of the Kuramoto transition (see Equation 4.37).

Equation 4.96 is valid for $R (x)^2 < \omega^2$, but we can formally extend the definition to the case where $R (x)^2 \geqslant \omega^2$ by using an appropriate density which is "delta-peaked" at the fixed point $\varphi^* (x)$. This yields

$$p_x (\varphi) = \begin{cases} \frac{\sqrt{\omega^2 - R(x)^2}}{2\pi |\omega - R(x) \sin[\varphi - \Theta(x) + \alpha]|} & \text{for } R (x)^2 < \omega^2 \\ \delta (\varphi - \varphi^* (x)) & \text{otherwise.} \end{cases} \tag{4.97}$$

With this notation, the time-independent version of the self consistency condition of Equation 4.91 can be written as

$$R (x) e^{i\Theta(x)} = \int_0^1 dx' \int_{-\pi}^{\pi} d\varphi \, p_{x'} (\varphi) G (x - x') e^{i\varphi} . \tag{4.98}$$

The structure of the right-hand side of this equation is now very similar to the right-hand side of Equation 4.35. However, instead of integrating over ω, the integration in Equation 4.98 is over x. Also, the weight function inside the integral is now $G(x - x')$ instead of $g(x)$. Accounting for the differences, but proceeding otherwise in an analogous way to the discussion of Section 4.1.5, one finally reaches

$$R(x)\,e^{i\Theta(x)} = ie^{-i\alpha} \int_0^1 dx'\,G(x-x')\,e^{i\Theta(x')}\frac{\omega - \sqrt{\omega^2 - R^2(x')}}{R(x')}. \qquad (4.99)$$

This equation is difficult to solve, since the unknown functions $R(x)$ and $\Theta(x)$ appear on both sides of it, and the parameter ω also needs to be chosen appropriately for a solution to exist. We also note that Equation 4.99 only fixes $\Theta(x)$ up to a constant shift. Therefore, in general, one needs to resort to numerical methods to calculate both $R(x)$ and $\Theta(x)$ iteratively. We also stress that, even if a self-consistent solution is found, it is not guaranteed to be stable.

One notable exception where Equation 4.99 can be solved to some extent analytically is the "cosine kernel"

$$G(x) = \frac{1}{2\pi}\left[1 + A\cos\left(\pi\left(x - \frac{1}{2}\right)\right)\right]. \qquad (4.100)$$

Inserting this kernel into Equation 4.99 shows that right-hand side has the structure

$$R(x)\,e^{i\Theta(x)} = c + a\cos\left(\pi\left(x - \frac{1}{2}\right)\right), \qquad (4.101)$$

where the coefficients a and c need to be self-consistently calculated. For further details we refer the reader to (Abrams and Strogatz 2006).

4.4 Bellerophon States

Recently, a novel coherent state has been discovered, occurring in globally coupled nonidentical oscillators in the proximity of the point where the transition to the synchronized behavior changes order, from continuous and reversible to first-order (i.e., occurring explosively, with a discontinuous and irreversible path). In this novel state, the oscillators form quantized clusters, where they are neither phase- nor frequency-locked. Their instantaneous speeds are different between clusters, but the instantaneous frequencies of the members of a given cluster form a characteristic cusped pattern and, more importantly, behave periodically in time so that their long-time averages are the same (Bi et al. 2016).

These states are intrinsically specular to the Chimera states, because they occur for globally coupled oscillators that, in addition, have to be as nonidentical as possible. Because of this specularity (and that of many of their fundamental properties), they were called *Bellerophon* states, as Bellerophon was the hero who, in Greek mythology, confronted with and eventually killed the monster Chimera. In the following, we will briefly describe their microscopic and macroscopic details.

The starting point is a Kuramoto-like model of N globally coupled phase oscillators obeying the equation

$$\dot{\vartheta}_i = \omega_i + \frac{\kappa|\omega_i|}{N}\sum_{j=1}^{N}\sin(\vartheta_j - \vartheta_i),\qquad(4.102)$$

where ϑ_i and ω_i are the instantaneous phase and the natural frequency of the ith member of the ensemble, respectively, and κ is the coupling strength. Once again, the level of synchronization can be measured by the order parameter

$$R = \frac{1}{N}\left\langle\left|\sum_{j=1}^{N}e^{i\vartheta_j}\right|\right\rangle_T,\qquad(4.103)$$

where the angled brackets denote time average.

Now, assume we prepare the system in specular conditions with respect to those for which Chimera states may emerge. In particular, the set of natural frequencies $\{\omega_i\}$ is drawn from a frequency distribution $g(\omega)$ that is even, symmetric and centered at 0. For example, one can choose

$$g(\omega) = \frac{\Delta}{2\pi}\left[\frac{1}{(\omega-\omega_0)^2 + \Delta^2} + \frac{1}{(\omega+\omega_0)^2 + \Delta^2}\right],\qquad(4.104)$$

that is, a bimodal Lorentz distribution, where Δ is the half width at half maximum of each constituting Lorentzian and $\pm\omega_0$ are their center frequencies. Note that $g(\omega)$ can be made unimodal by choosing parameters such that $\omega_0/\Delta \leqslant \sqrt{3}/3$.

Since only the ratio ω_0/Δ matters, one can take $\Delta = 1$ and let ω_0 progressively increase, leading to a gradual increase of the distance between the two peaks of $g(\omega)$. The result is that, for small values of ω_0, an irreversible, first-order-like, abrupt transition is observed, with a characteristic hysteresis area. Bi et al. (2016) showed that the width of the hysteresis progressively shrinks as ω_0 increases, leading to an eventual crossover to a reversible, second-order-like, continuous transition for $\omega \approx 1.7$.

As long as $g(\omega)$ is symmetric, the critical point for the backward transition is $\kappa_b = 2$ (Hu et al. 2014). However, the critical point κ_f for the forward transition varies with ω_0 according to Bi et al. (2016)

$$\kappa_f = \frac{4}{\sqrt{1+(\omega_0/\Delta)^2}}.\qquad(4.105)$$

This equation indicates that the critical point for the forward transition is uniquely determined by the dimensionless parameter ω_0/Δ. In particular, when $\omega_0 = 0$, $g(\omega)$ degenerates into the typical unimodal Lorentzian distribution, and Equation 4.105 predicts $\kappa_f = 4$.

The expression also explains the switch from first- to second-order-like transition. When $\omega_0/\Delta = \sqrt{3}$, $\kappa_f = \kappa_b = 2$, and the forward and backward transition points coincide. However, the hysteresis area does not immediately disappear. In fact, at $\omega_0/\Delta = \sqrt{3}$, a Hopf bifurcation occurs during both the forward and backward processes, and both bifurcations are continuous. For the forward direction, the system first undergoes a continuous transition, followed by an explosive synchronization (ES) transition, and a similar scenario of transitions characterizes the backward direction. Thus, an initial parameter range $\omega_0/\Delta > \sqrt{3}$ exists, where the system undergoes the cascade of one continuous and one explosive transition for both forward and backward directions. A further increase of ω_0/Δ causes the hysteresis area to actually disappear, leading to a situation where only continuous transitions occur in the system. It is in this latter regime, i.e., close to a tricritical point in parameter space, that the Bellerophon states emerge.

For the sake of example, one can take the case of $\omega_0/\Delta = 3$, where the system exhibits two continuous transitions at $\kappa_f = 4/\sqrt{10} \approx 1.26$ and $\kappa_b = 2$, respectively. Here, three parameter regimes can be identified: (i) $\kappa < \kappa_f$, (ii) $\kappa_f < \kappa < \kappa_b$, and (iii) $\kappa > \kappa_b$. In regime (i), the coupling strength is small and the system is in the trivial incoherent state. In regime (iii), the coupling is so strong that the system goes into the fully synchronized state, in which all oscillators split into fully synchronized clusters. Bellerophon phases are steady states occurring in the middle regime (ii), i.e., during the path to full synchronization. The interested reader is here referred to (Bi et al. 2016) for a full description of the main features of these states, which are characterized by three quantities: the instantaneous phases ϑ_i, the instantaneous (angular) speed $\dot{\vartheta}_i$, and the average speed $\langle \dot{\vartheta}_i \rangle$.

4.5 Oscillation Quenching

In many applications, the coupling strength between individual oscillators can change over time, either through inherent parameter drift or due to external manipulation. It is important to understand what the consequences of such changes are. Let us, for example, consider the pacemaker cells in the heart. Because of natural fluctuations, the frequencies of individual cells will in general be different. However, if the interaction between cells is strong enough, a common rhythm will emerge, which causes the coherent contraction of the heart. Drugs or other factors can influence the strength of the interaction between cells, and one would like to predict how this affects the functioning of the heart.

For small coupling, individual oscillators behave as if they were uncoupled and oscillate incoherently. As the coupling is increased a counterintuitive effect is observed in many oscillator models: the motion of the system comes to a complete stop at intermediate coupling. This phenomenon is called *oscillation quenching*, and was discovered by Yamaguchi and Shimizu (1984). For larger coupling, a state of synchronized oscillation is often recovered.

We distinguish two manifestations of oscillation quenching: amplitude death (AD) and oscillation death (OD). Historically, the two term OD and AD were used almost synonymously, but more recently a useful terminology has been established (Koseska et al. 2013a,b; Zou et al. 2013) that defines them as two clearly distinct phenomena. The characteristic feature of AD is that the coupling stabilizes a fixed point of the oscillator that already exists without coupling. Typically, this fixed point is the origin, and therefore AD is associated with the stabilization of the homogeneous steady state of all individual oscillators residing at the origin. This is in contrast to OD, where a stable inhomogeneous steady state emerges, i.e., a fixed point where not all oscillators reside at the origin.

More formally, and using the definitions of Section 4.2.2, consider a system of N coupled oscillators

$$\dot{\mathbf{u}}_j(t) = \mathbf{f}_j(\mathbf{u}_j) + \sum_{\substack{k=1 \\ k \neq j}}^{N} \mathbf{g}_{jk}(\mathbf{u}_j, \mathbf{u}_k) \ . \tag{4.106}$$

The system is said to show oscillation quenching if a stable fixed point

$$\mathbf{U}^* = (\mathbf{u}_1^*, \ldots, \mathbf{u}_N^*) \tag{4.107}$$

exists. In this case, we call the fixed point AD if the condition $\mathbf{f}_j(\mathbf{u}_j^*) = \mathbf{0}$ is fulfilled for all j. If this condition is violated for at least one j, then we call the fixed point OD.

Oscillation quenching is in clear contrast to the situation in the Kuramoto model, where the transition to synchrony was explained by the continuous growth a synchronized cluster. Since the Kuramoto model is symmetric under a simultaneous change in the phases of all oscillators, it does not possess an isolated stable fixed point, and therefore oscillation quenching is not possible.

4.5.1 Amplitude Death

AD is only possible if the local dynamics possess a fixed point, and therefore by definition it cannot occur in a simple phase oscillator model with local dynamics

$f_j(\vartheta) \neq 0$. AD can therefore only occur in systems where the oscillators are at least two-dimensional. From now on let us assume that the origin $\mathbf{0}$ is the only fixed point of the local dynamics.

Using the AD condition in Equation 4.106 shows that AD is only possible if

$$\sum_{\substack{k=1 \\ k \neq j}}^{N} \mathbf{g}_{jk}(\mathbf{0}, \mathbf{0}) = \mathbf{0} \quad \text{for all } j. \tag{4.108}$$

A natural way of satisfying this condition is the use of diffusive coupling, in which case Equation 4.108 is automatically fulfilled.

A simple model that illustrates the phenomenon of AD is the *Aizawa model* by Aizawa (1976); Mirollo and Strogatz (1990),

$$\dot{z}_k = \left(1 - |z_k|^2 + i\omega_k\right) z_k + \frac{K}{N} \sum_{j=1}^{N} \left(z_j - z_k\right). \tag{4.109}$$

Here, $z_k \in \mathbb{C}$ are complex dynamical variables, and K is a real coupling coefficient. The frequencies $\omega_k \in \mathbb{R}$ are randomly chosen from a Gaussian distribution with standard deviation σ. The individual oscillators are canonical limit-cycle oscillators as in Equation 4.1 with the choice of $\alpha = \beta = 1$. Without interaction ($K = 0$), the individual oscillators follows a stable limit cycle of period $2\pi/\omega_k$ and amplitude 1. Each oscillator also features an unstable fixed point at the origin. In difference from Equation 4.3, where direct coupling was used, the coupling in Equation 4.109 is diffusive.

Let us study the behavior of the average amplitude of the oscillators

$$\langle |z| \rangle = \frac{1}{N} \sum_{k=1}^{N} |z_k| \tag{4.110}$$

after a stationary state has been reached. This is shown for various coupling strengths in Figure 4.19 in the case of $\sigma = 2$. As expected, one finds $\langle |z| \rangle = 1$ for the unsynchronized case at vanishing coupling. Also, for large coupling, $\langle |z| \rangle$ approaches 1, which is consistent with a transition towards a fully synchronized case. What is surprising, however, is that $\langle |z| \rangle$ completely vanishes for a large interval of intermediate couplings. In this regime the fixed point at the origin is stable and the individual oscillators are away from their natural limit cycle.

To understand this phenomenon analytically, consider the Jacobian matrix of system (4.109) at the origin (Mirollo and Strogatz 1990), which is given by

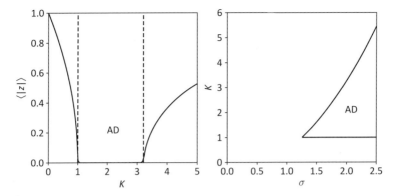

Figure 4.19 Left panel: average amplitude according to Equation 4.110 of system in Equation 4.109 at time $t = 200$. Parameters are $N = 2048$ and $\sigma = 2$. Right panel: region of amplitude death in σ-K parameter plane.

$$
\mathbf{J} = \begin{pmatrix} 1 - K + i\omega_1 & 0 & \cdots & & 0 \\ 0 & \ddots & \ddots & & \vdots \\ \vdots & & \ddots & \ddots & 0 \\ 0 & & \cdots & 0 & 1 - K + i\omega_N \end{pmatrix} + \frac{K}{N} \begin{pmatrix} 1 & \cdots & 1 \\ \vdots & \ddots & \vdots \\ 1 & \cdots & 1 \end{pmatrix}. \tag{4.111}
$$

The fixed point at the origin is stable if none of the eigenvalues of $\mathbf{J}$ has a positive real part. A necessary condition for this to occur is that $\Re\,(\mathrm{Tr}\mathbf{J}) \leqslant 0$, which implies that

$$
N\,(1 - K) + K \leqslant 0, \tag{4.112}
$$

and consequently

$$
K \geqslant \frac{N}{N-1} > 1. \tag{4.113}
$$

The specific form of $\mathbf{J}$ allows us to calculate its characteristic polynomial analytically. To simplify the notation, let us introduce the matrix $\mathbf{A} = \mathbf{J} - \lambda \mathbb{1}$ and the variables

$$
\mu_k = 1 - K + i\omega_k - \lambda. \tag{4.114}
$$

The matrix $\mathbf{A}$ is then explicitly of the simple form

$$
A_{kj} = \begin{cases} \mu_k + \frac{K}{N} & \text{for } k = j \\ \frac{K}{N} & \text{otherwise.} \end{cases} \tag{4.115}
$$

The characteristic polynomial for a system with N oscillators is given by $\chi_N = \det \mathbf{A}$. Denoting by $\mathbf{M}_{jk}$ the minors of $\mathbf{A}$, one obtains by a standard expansion along the first row

$$\chi_N = \det \mathbf{A} = \sum_{k=1}^{N} (-1)^{k+1} A_{1k} \det \mathbf{M}_{1k}$$

$$= \mu_1 \det \mathbf{M}_{11} + \det \begin{pmatrix} \frac{K}{N} & \frac{K}{N} & \cdots & & \frac{K}{N} \\ \frac{K}{N} & \frac{K}{N} + \mu_2 & \ddots & & \vdots \\ \vdots & & \ddots & \ddots & \frac{K}{N} \\ \frac{K}{N} & \cdots & & \frac{K}{N} & \frac{K}{N} + \mu_N \end{pmatrix} \qquad (4.116)$$

$$= \mu_1 \chi_{N-1} + \frac{K}{N} \prod_{k=2}^{N} \mu_k ,$$

where, in the last step, we have used the fact that $\det \mathbf{M}_{11} = \chi_{N-1}(\lambda)$ is simply the characteristic polynomial of a system of dimension $N - 1$, with oscillator $k = 1$ removed. To obtain the second term in the last line, note that the determinant in the preceding line is not affected if the first row is subtracted from all subsequent rows, as it is the product of the remaining diagonal elements. Thus, Equation 4.116 expresses χ_N in terms of χ_{N-1}. Repeating this process, and finally substituting Equation 4.114, leads to the following expression for the characteristic polynomial:

$$\chi_N(\lambda) = \left[\prod_{k=1}^{N} (1 - K + i\omega_k - \lambda) \right] \left(1 + \frac{K}{N} \sum_{j=1}^{N} \frac{1}{1 - K + i\omega_k - \lambda} \right). \qquad (4.117)$$

The AD fixed point is stable if all roots of Equation 4.117 have nonpositive real part. This is equivalent to the condition that no solution of the complex equation

$$1 = \frac{K}{N} \sum_{j=1}^{N} \frac{1}{\lambda - 1 + K - i\omega_k} \qquad (4.118)$$

has positive real part. Once the frequencies ω_k are known explicitly, it is possible to check this condition numerically.

Our main interest is in the limit of infinitely many oscillators, where the ω_k are independent random variables with a given probability distribution $g(\omega)$. In this limit it is then possible to replace the sum in Equation 4.118 with a weighted integral:

$$1 = K \int_{-\infty}^{\infty} \frac{g(\omega)\, d\omega}{\lambda - 1 + K - i\omega} . \qquad (4.119)$$

As in the finite N case, the AD solution is stable with probability going to as N goes to infinity if Equation 4.119 does not allow for a solution λ with positive real part.

Finding and checking all roots of Equation 4.119 is in general difficult. However, in the common case where $g(\omega)$ is even and nonincreasing for $\omega > 0$, Mirollo and Strogatz (1990) showed that the stability of the solution can be conveniently assessed by evaluating the right-hand side of Equation 4.119 for $\lambda = 0$. One finds that AD is stable if the conditions $K > 1$ and

$$1 > K \int_{-\infty}^{\infty} \frac{g(\omega)\, d\omega}{K - 1 - i\omega} = 2K \int_{0}^{\infty} \frac{K - 1}{(K - 1)^2 + \omega^2} g(\omega)\, d\omega \qquad (4.120)$$

are fulfilled.

These conditions can be used to determine the stability region of the AD state, shown in the right-hand panel of Figure 4.19 for a Gaussian distribution with standard deviation σ. We see that the AD regime only exists for sufficiently large σ, and then grows quickly in size. Because of condition Equation 4.113, AD is a strong coupling effect, i.e., it requires $K > 1$.

We have focused in this section on the Aizawa model Equation 4.109, which is paradigmatic for many similar models with nonidentical oscillators and instantaneous homogeneous and diffusive couplings. In general terms, the propensity for AD increases with the inhomogeneity of the oscillators, and for simple diffusive coupling AD does not exist for an ensemble of identical oscillators.

However, if we allow for more general couplings, AD is also possible in systems of identical oscillators. In particular, time-delayed coupling (Ramana Reddy et al. 1998) or coupling between dissimilar variables (Karnatak et al. 2007) have been shown to be efficient mechanisms for this purpose. More recently, it was shown that AD is also possible in ensembles of identical two-dimensional time-delayed oscillators with direct coupling. Further examples of AD can be found in recent reviews (Saxena et al. 2012; Koseska et al. 2013a).

4.5.2 Oscillation Death

The second mechanism for oscillation quenching in a system of coupled oscillators is OD. According to our definition, it requires that the condition

$$\mathbf{f}_j\left(\mathbf{u}_j^*\right) = -\sum_{\substack{k=1 \\ k \neq j}}^{N} \mathbf{g}_{jk}\left(\mathbf{u}_j^*, \mathbf{u}_k^*\right) \neq \mathbf{0} \qquad (4.121)$$

is fulfilled at least for one j. The OD fixed point is induced by the coupling terms, and for its existence it is not even required that the isolated local dynamics possess a fixed point.

A simple example for this phenomenon is given by a system of two identical phase oscillators:

$$
\begin{aligned}
\dot{\varphi}_1 &= \omega - A \sin(\varphi_1) + K \sin(\varphi_1) \sin(\varphi_1 - \varphi_2) \\
\dot{\varphi}_2 &= \omega - A \sin(\varphi_2) + K \sin(\varphi_2) \sin(\varphi_2 - \varphi_1),
\end{aligned}
\qquad (4.122)
$$

where A and K are real and nonnegative parameters. The local dynamics $f_1(\varphi) = f_2(\varphi) = \omega - A\sin(\varphi_1)$ do not possess a fixed point if $\omega > A$. The system is invariant under exchange of the two oscillators $(\varphi_1, \varphi_2) \rightarrow (\varphi_2, \varphi_1)$ and the coupling vanishes for $\varphi_1 = \varphi_2$. However, it is important to note that the system is not invariant under simultaneous shifts of both phase variables $(\varphi_1, \varphi_2) \rightarrow (\varphi_1 + \Delta\varphi, \varphi_2 + \Delta\varphi)$, as was the case for the Kuramoto model. This broken translational symmetry allows for the emergence of isolated fixed points.

The different dynamical regimes of Equation 4.122 in K–A parameter space are shown in Figure 4.20. One observes that for sufficiently large K a pair of symmetry-broken stable fixed points of the form (φ_1, φ_2) and (φ_2, φ_1) with $\varphi_1 \neq \varphi_2$ appears through a saddle node bifurcation.

The inherent feature of symmetry-broken fixed points, which is evident in the simple system of Equation 4.122, persists as we increase the number of oscillators, or exchange the individual oscillators by more complicated ones. This naturally leads to systems with a high degree of multistability. Inspired by the ideas of Turing (1952), it has been suggested that OD is a mechanism behind pattern formation in biology (Koseska et al. 2009, 2010a) and physics (Kuntsevich and Pisarchik, 2001).

Interestingly, both AD and OD can appear together in simple systems. Let us for example consider a variant of the Aizawa model for $N = 2$ (Koseska et al. 2013b),

$$\dot{z}_1 = \left(1 - |z_1|^2 + i\omega_1\right)z_1 + \varepsilon\Re(z_2 - z_1),$$
$$\dot{z}_2 = \left(1 - |z_2|^2 + i\omega_2\right)z_2 + \varepsilon\Re(z_1 - z_2),$$
(4.123)

where K, ω_1 and ω_2 are real parameters. In comparison with the original Aizawa model, the appearance of the real part in the coupling now breaks the phase shift

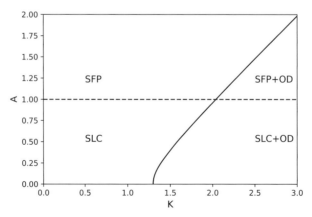

Figure 4.20 Regimes of stable solutions for the model of Equation 4.122 for $\omega = 1$. SLC indicates a synchronized limit cycle with $\varphi_1(t) = \varphi_2(t)$, SFP: a stable fixed point at $\varphi_1 = \varphi_2 = \arcsin(1/A)$; OD: pair of symmetry-broken stable fixed points. Solid and dashed lines are saddle node bifurcation lines for OD and SFP states, calculated with AUTO07p (Doedel et al. 2007).

symmetry $(\varphi_1, \varphi_2) \rightarrow (\varphi_1 + \Delta\varphi, \varphi_2 + \Delta\varphi)$. This change allows for the emergence of a pair of nontrivial fixed points, which in the case of $\omega_1 = \omega_2 = \omega$ are given by (Koseska et al. 2013b)

$$z_\pm^* = -\frac{\omega y_\pm^*}{\omega^2 + 2\varepsilon y_\pm^*} + i y_\pm^*, \qquad (4.124)$$

with

$$y_\pm^* = \sqrt{\frac{\varepsilon - \omega^2 \pm \sqrt{\varepsilon^2 - \omega^2}}{2\varepsilon}}. \qquad (4.125)$$

The nontrivial fixed points are generated at $2\varepsilon = \omega^2 + 1$ through a pitchfork bifurcation, but are not necessarily stable.

A full bifurcation analysis of Equation 4.123 for $\omega_1 \neq \omega_2$ cannot be performed analytically, and requires the use of numerical methods to follow individual solution branches. The result of such an analysis is shown in Figure 4.21, which illustrates the dynamical regimes in an $\varepsilon - \Delta$ diagram, where $\Delta = \omega_1/\omega_2$. It is interesting to note that if the pitchfork bifurcation discussed before is continued for large Δ, then it regulates a direct transition from an AD state to and OD state (close to vertical line in Figure 4.21).

4.5.3 Chimera Death

At first sight, the two phenomena of OD and chimera states seem very different, since chimera states involve the coexistence between incoherent and coherent solutions, while OD involves the emergence of inhomogeneous fixed points.

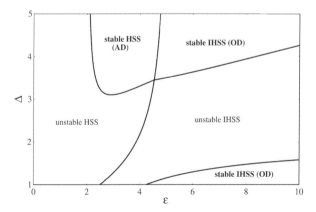

Figure 4.21 Dynamical regimes of coupled Stuart–Landau oscillators Equation 4.123, with $\omega_2 = 2$ and $\omega_1 = \omega_2\Delta$. HSS (IHSS) denotes the homogeneous (inhomogeneous) steady state. Reprinted with permission from Koseska et al. (2010b).

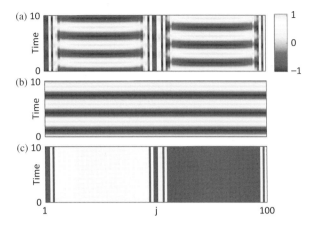

Figure 4.22 Space-time plots of evolution of $\Im z_k$ according to Equation 4.126 for $K = 14$, $N = 100$ and $\omega = 2$. (a) $P = 4$ (a); (b) $P = 5$ (c) $P = 33$. Reprinted with permission from Zakharova et al. (2014).

However, on an abstract level, both concepts are related to spontaneous symmetry breaking, and the emergence of patterns from a homogeneous state.

This idea was explored by Zakharova et al. (2014) and an interesting connection was discovered in the system of N equations

$$\dot{z}_j = \left(1 - |z_j|^2 + i\omega\right) z_j + \frac{K}{2P} \sum_{k=j-P}^{j+P} \Re\left(z_k - z_j\right), \tag{4.126}$$

where $K < N$ is an integer number and K and ω are real parameters. A ring topology is assumed, i.e., $z_j = z_{j+N}$. This equation is similar to the classical discrete chimera Equation 4.72. However, the phase oscillators are now replaced by Stuart–Landau oscillators. In addition, similarly to Equation 4.123, the appearance of the real part breaks the rotational symmetry of the model. With this changes it is now possible to "freeze" a chimera state in time. This phenomenon, called "chimera death," is shown in Figure 4.22(c), where two large uniform regions are separated by regions with random changes of phases between neighboring sites. This chimera death state is a result of large coupling range ($P = 33$) and strong coupling ($\sigma = 14$). For smaller values of P, traditional chimera states (Figure 4.22(a)) and locked states (Figure 4.22(b)) are observed.

4.6 Auto-Synchronization and Time-Delayed Feedback

So far, we have studied a number of different occurrences of synchronization, all of which involve the interaction between at least two systems. It might therefore be surprising to realize that a system can also synchronize with itself, if we allow for a *time-delay* coupling. Consider, for example, a time-delay equation of the form

$$\dot{\mathbf{x}}(t) = \mathbf{f}(\mathbf{x}(t)) + \mathbf{K}(\mathbf{x}(t - \tau_d), \mathbf{x}(t)). \tag{4.127}$$

Here, $\mathbf{x}(t) \in \mathbb{R}^n$ is the state vector of the system at time t and $\mathbf{f}(\mathbf{x})$ describes the dynamics of the system without coupling. The coupling term is a vector function that depends on the current state vector $\mathbf{x}(t)$ as well as on the time-delayed state vector $\mathbf{x}(t - \tau_d)$ of the *same* system. This is in contrast to the case of lag synchronization considered in Section 2.6, where the time delay state of a second system appeared. The *delay time* τ_d is a real parameter. Depending on the system under consideration, the physical origin of the delay term might be due to a finite signal propagation time (such as in an optical system using a mirror), a finite processing time or a finite response time of an actuator or sensor. In some physical systems, the delay term is unavoidable, while in other cases it might be deliberately added to obtain desired dynamics. In particular, as we discuss below in more detail, a proper choice of the coupling term can cause the system to synchronize with a version of itself that is shifted in time. This effect is sometimes called *auto-synchronization*.

Before we study examples and applications of time delay equations, it is necessary to point out an important mathematical fact about the dimensionality of their phase space. If we want to employ Equation 4.127 to calculate the evolution of $\mathbf{x}(t)$ for $t > 0$, then it is not sufficient to provide only the initial condition $\mathbf{x}_0 = \mathbf{x}(0)$ at a single point in time $t = 0$. Instead, we require knowledge of $\mathbf{x}(t)$ in the full time interval from $-\tau$ up to 0. In other words, instead of a single initial point, we now require a (sufficiently smooth) *initial function* $\mathbf{x}_0(t)$, defined for $t \in [-\tau_d, 0]$. This means, however, that the space of initial function is infinite-dimensional, and therefore the phase space of the time-delay equation 4.127 is formally also infinite-dimensional.

4.6.1 Chaos Control

One major application of time delay equations of the form Equation 4.127 is the control of nonlinear and chaotic systems. While nonlinear systems are ubiquitous and often feature chaotic behavior, it is desirable to actually avoid the chaotic state in many applications. The question therefore arises as to whether we can influence a given chaotic system in such a way that the chaotic state is suppressed, and a desired periodic behavior emerges.

The key observation is that there are in general infinitely many unstable periodic orbits embedded in a chaotic attractor. More precisely, for any ball of arbitrarily small size ε around any point in a chaotic attractor, there exists a periodic orbit going through that ball (Ott et al. 1990; Grebogi et al. 1988). It would thus be possible to control a chaotic system, if one of these existing unstable orbits could be transformed into a stable one. This idea of "controlling chaos" was first implemented by Ott et al. (1990). However, the method used involves a fairly complicated construction of the control force, which is difficult to implement in practical systems, and in general requires the use of a computer.

Only a short time later, Pyragas (1992) demonstrated that a deceptively simple method can stabilize unstable periodic orbits in a chaotic attractor very efficiently. The appeal of this method is that it requires little prior knowledge about the system to be controlled and the desired target orbit. Only a single scalar parameter, namely the period τ_p of the target periodic orbit, needs to be known with sufficient precision. In addition, coupling only to a single scalar dynamical variable of the system, even if the system itself is high-dimensional, is often sufficient.

The method works as follows. Consider a chaotic system

$$\dot{\mathbf{x}}(t) - \mathbf{f}(\mathbf{x}(t)), \tag{4.128}$$

and assume that there exists a particular periodic orbit

$$\mathbf{x}^*(t) = \mathbf{x}^*\left(t - \tau_p\right) \tag{4.129}$$

of period τ_p that we wish to stabilize. To achieve control, Pyragas suggested the modification of the original dynamics to a time delay feedback scheme

$$\dot{\mathbf{x}}(t) = \mathbf{f}(\mathbf{x}(t)) + \mathbf{K}(\mathbf{x}(t - \tau_d) - \mathbf{x}(t)), \tag{4.130}$$

where $\mathbf{K}$ is an $n \times n$ matrix. A schematic view of this equation in the form of a block diagram is shown in Figure 4.23. If we choose the delay time τ_d to be equal to the period of the periodic orbit τ_p, from Equation 4.129 it is clear that, if the system evolves on the desired orbit, the coupling term identically vanishes for any $\mathbf{K}$. This means that the periodic orbit is a solution of Equation 4.130 whose existence is not affected by the coupling term. The important point here is that the stability of the periodic orbit *is* affected by $\mathbf{K}$, and indeed it is possible to stabilize

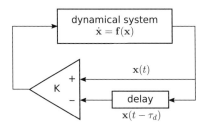

Figure 4.23 Schematic view of the time-delayed feedback method Equation 4.130.

a previously unstable orbit $\mathbf{x}^*(t)$ with a suitable choice of $\mathbf{K}$. This method is often called *Pyragas control*. For a broader introduction into the interesting field of chaos control, we refer the reader to Schöll and Schuster (2008).

4.6.2 Experimental Realization

One major advantage of Pyragas control is that it can be implemented without relying on a computer to calculate the control force. The delay term can be obtained for example through a delay line in an electrical system, as demonstrated experimentally by Pyragas and Tamaševičius (1993). Their experimental setup, shown in Figure 4.24, uses a nonlinear electronic oscillator that delivers a chaotic output signal without control. The control circuit consists of a delay line that delays the signal by a time corresponding to the period of the desired periodic orbit. The difference between the original signal and the delayed one is amplified and fed back into the nonlinear circuit. From the time traces in the right-hand panel of Figure 4.24 we see that the chaotic signal becomes periodic after control is switched on. When control is successful, the delayed and original signals precisely match, and as a result the actual control signal vanishes. This vanishing of the control force upon successful control is often called *noninvasive* control.

Since this initial experimental realization, the time delay feedback scheme has been successfully applied in many different contexts and was found useful in technological applications. For example, time delay feedback been applied to control the cantilever oscillations in atomic force microscopy (Yamasue et al. 2009), thereby improving the image quality. Apart from the already mentioned electronic and mechanical implementations, time-delayed feedback control has also been experimentally employed in optical (Arecchi et al. 2002; Schikora et al. 2006) and

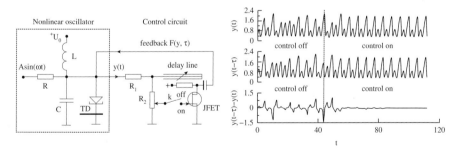

Figure 4.24 Left: electronic nonlinear oscillator with control circuit. Right: time traces of output voltage (top), delayed output voltage (middle) and applied control signal (bottom). Reprinted with permission from Pyragas and Tamaševičius (1993).

chemical (Bertram et al. 2003) systems, and is therefore a truly interdisciplinary technique.

Another interesting experimental application of time-delayed feedback is to stabilize periodic orbits while some of the system parameters are adiabatically changed. In particular, this allows one to stabilize a periodic orbit across a bifurcation and into a parameter regime where it would be unstable in the original system (Kiss et al. 2006; Sieber et al. 2008). In this case, the underlying system is not necessarily chaotic, and the control scheme serves to stabilize the desired orbit even in the presence of competing stable equilibria or limit cycles.

4.6.3 Theory

The main problem in the theory of Pyragas control is to design a suitable coupling matrix $\mathbf{K}$ to enable control of a given unstable periodic orbit. It is therefore necessary to assess the linear stability of the controlled system, Equation 4.130.

Let us assume that $\tau_d = \tau_p$. The stabilization scheme Equation 4.130 is successful if a small perturbation $\delta\mathbf{x}(t)$ added to the periodic orbit $\mathbf{x}^*(t)$ does not grow exponentially in time. Since we are dealing with a time delay system, the perturbation $\delta\mathbf{x}(t)$ is also defined on some interval $t \in [-\tau_d, 0]$, and therefore the space of all possible perturbations is an infinite-dimensional vector space. By inserting

$$\mathbf{x}(t) = \mathbf{x}^*(t) + \delta\mathbf{x}(t) \tag{4.131}$$

into Equation 4.130, to first order in $\delta\mathbf{x}(t)$ one obtains the evolution equation

$$\delta\dot{\mathbf{x}}(t) = J\mathbf{f}\left(\mathbf{x}^*(t)\right)\delta\mathbf{x}(t) + \mathbf{K}\left(\delta\mathbf{x}\left(t - \tau_p\right) - \delta\mathbf{x}(t)\right), \tag{4.132}$$

where $J\mathbf{f}(\mathbf{x}^*(t))$ is the Jacobian matrix of $\mathbf{f}$ evaluated at the point $\mathbf{x}^*(t)$.

Floquet Multipliers and Fundamental Matrix

Let us first consider the uncontrolled case of $\mathbf{K} = \mathbf{0}$. It is useful to introduce a time-dependent $n \times n$ matrix $\Phi(t)$ that fulfils the dynamical matrix equation

$$\dot{\Phi}(t) = J\mathbf{f}\left(\mathbf{x}^*(t)\right)\Phi(t) \tag{4.133}$$

with initial condition $\Phi(0) = \mathbb{1}$. This matrix is often called the *fundamental matrix*, and by comparing with Equation 4.132, it is clear that $\delta\mathbf{x}(t) = \Phi(t)\delta\mathbf{x}(0)$ in the case of $\mathbf{K} = \mathbf{0}$. The algebraic eigenvalues $\{\mu_1, \ldots, \mu_n\}$ of the matrix $\Phi(\tau_p)$ are called the *Floquet multipliers* of the periodic orbit, and they are in general complex numbers. The periodic orbit is unstable if at least one of the Floquet multipliers fulfils the condition of $|\mu_k| > 1$. We furthermore denote by $\mathbf{v}_1^0, \ldots, \mathbf{v}_n^0$ the generalized (Jordan) eigenvectors of $\Phi(\tau_p)$.

In the case of an autonomous system, one observes that for small δt

$$\mathbf{x}^* (t + \delta t) - \mathbf{x}^* (t) \approx \delta t \dot{\mathbf{x}}^* (t) = \delta t \mathbf{f} \left(\mathbf{x}^* (t) \right) = \delta t \mathbf{f} \left(\mathbf{x}^* \left(t + \tau_p \right) \right) . \tag{4.134}$$

However, a small perturbation $\delta \mathbf{x} (t) = \delta t \mathbf{f} (\mathbf{x}^* (t)) = \delta \mathbf{x} (t + \tau_p)$ evidently fulfils Equation 4.132 for $\mathbf{K} = \mathbf{0}$, and therefore we can conclude that one of the Floquet multipliers equals 1 and that the vector $\mathbf{f} (\mathbf{x}^* (t))$ is the associated eigenvector. Without loss of generality, we choose $\mu_1 = 1$ and $\mathbf{v}_1^0 = \mathbf{f} (\mathbf{x}^* (t))$.

Diagonal Coupling

The discussion of stability is significantly simplified if the matrix $\mathbf{K} = \kappa \mathbb{1}$ is a multiple of the identity matrix. In this case, which we call *diagonal coupling*, $\mathbf{K}$ commutes with $J\mathbf{f} (\mathbf{x}^* (t))$ and we can insert the Ansatz

$$\delta \mathbf{x} (t) = g (t) \, \Phi (t) \, \mathbf{v}_k^0, \tag{4.135}$$

with $g (0) = 1$ into Equation 4.132 to yield

$$\dot{g} (t) \, \Phi (t) \, \mathbf{v}_k^0 + g (t) \, \dot{\Phi} (t) \, \mathbf{v}_k^0 = J\mathbf{f} \left(\mathbf{x}^* (t) \right) \delta \mathbf{x} (t)$$
$$+ \mathbf{K} \left(g (t - \tau_d) \, \Phi (t - \tau_d) - g (t) \, \Phi (t) \right) \mathbf{v}_k^0 , \tag{4.136}$$

hence

$$\dot{g} (t) \, \Phi (t) \, \mathbf{v}_k^0 = \mathbf{K} \left(g (t - \tau_d) \, \mu_k^{-1} - g (t) \right) \Phi (t) \, \mathbf{v}_k^0 , \tag{4.137}$$

and

$$\dot{g} (t) = \mathbf{K} \left(g (t - \tau_d) \, \mu_k^{-1} - g (t) \right) . \tag{4.138}$$

The last equation is a linear time-delay differential equation in one dimension (Amann et al. 2007). Except for the original Floquet multiplier μ_k all details of the system have dropped out. To solve this equation, we choose $g (t) = e^{\Lambda t}$ and arrive at the characteristic equation

$$\Lambda = \mathbf{K} \left(e^{-\Lambda \tau_d} \mu_k^{-1} - 1 \right) . \tag{4.139}$$

The difficulty in solving Equation 4.139 is that the parameter Λ appears inside the exponent, and it therefore has potentially infinitely many solutions. Since we are ultimately interested in the stability of the controlled system, from Equation 4.135 we obtain

$$\delta \mathbf{x} \left(\tau_p \right) = e^{\Lambda \tau_p} \mu_k \mathbf{v}_k^0 . \tag{4.140}$$

Stability of the controlled orbit is therefore achieved if the condition

$$\left| e^{\Lambda \tau_p} \mu_k \right| \leqslant 1 \tag{4.141}$$

holds for all k and all solutions Λ of Equation 4.139.

In order to solve Equation 4.139 explicitly, it is useful to introduce the *Lambert function* $W(x)$, which is defined as the solution of the equation

$$W(x)\, e^{W(x)} = x \,. \tag{4.142}$$

This equation has in general many solutions, which are often indicated by indexing the Lambert function as $W_k(x)$ where k runs from $-\infty$ to $+\infty$. By convention, the branch with the largest real part is denoted by W_0.

Let us manipulate Equation 4.139 into the form

$$(\Lambda + \mathbf{K})\, \tau_d e(\Lambda + \mathbf{K})\, \tau_d = \tau_d \mathbf{K} \mu_k^{-1} e^{\mathbf{K}\tau_d} \,. \tag{4.143}$$

Comparing with Equation 4.142 one identifies

$$x = \tau_d \mathbf{K} \mu_k^{-1} e^{\mathbf{K}\tau_d} \tag{4.144}$$

and

$$W(x) = (\Lambda + \mathbf{K})\, \tau_d \,, \tag{4.145}$$

and thus obtains

$$(\Lambda + \mathbf{K})\, \tau_d = W\left(\tau_d \mathbf{K} \mu_k^{-1} e^{\mathbf{K}\tau_d}\right) \,. \tag{4.146}$$

Exponentiation on both sides and using the fact that $e^{W(x)} = x/W(x)$ then gives

$$\mu_k e^{\Lambda \tau_d} = \frac{\tau_d \mathbf{K}}{W\left(\tau_d \mathbf{K} \mu_k^{-1} e^{\mathbf{K}\tau_d}\right)} \,, \tag{4.147}$$

and it is possible to check if, for a given $\mathbf{K}$, the condition of Equation 4.141 is fulfilled for all branches of the Lambert function.

Odd Number Limitation

In the case where the coupling is not diagonal, it is in general not possible to find analytical stability criteria. Nevertheless, one would like to obtain some guidance on the design of the control matrix K, since it is very difficult to find a suitable coupling matrix by chance, especially if the dimension of the system is large. Unfortunately, no sufficient condition for stability is known. However, there has recently been some progress in finding a number of useful necessary conditions, which are required for Pyragas control to be successful (Hooton and Amann 2012).

The criterion formulated by Hooton and Amann (2012) requires the knowledge of the number of Floquet multipliers of the uncontrolled orbit $\mathbf{x}^*(t)$ that are real and larger than unity. Assume that there are m multipliers with $\mu_k > 1$.

Let us again consider the control scheme of Equation 4.141, but now let us explicitly allow for $\tau_p \neq \tau_d$. Let us denote by $\tau_i(\tau_d)$ the period of the orbit that is induced by the control force with a given delay time τ_d. We know that

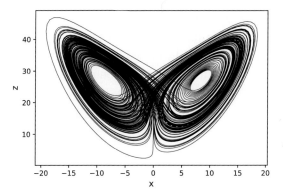

Figure 4.25 Successful control of the primary orbit of the Lorenz system Equation 4.149 by Pyragas control scheme Equation 4.130 using the control matrix in Equation 4.150 with $\kappa = 1$ and $\tau_d = 1.558652$. The controlled orbit is shown as dashed thick line, embedded in the chaotic Lorenz attractor (thin solid line).

$\tau_i\left(\tau_d = \tau_p\right) = \tau_p$, since the periodic orbit is by construction a solution of Equation 4.141 for $\tau_d = \tau_p$. For τ_d close to τ_p this periodic orbit is in general not immediately destroyed. It is therefore possible to define $\tau_i\left(\tau_d\right)$ and its derivative $\tau_i'\left(\tau_d\right)$ with respect to τ_d at least for some small interval around τ_p.

The odd number limitation then states that a necessary condition for a periodic orbit under the control scheme Equation 4.141 to be stable is that

$$(-1)^m \left(1 - \tau_i'\left(\tau_p\right)\right) > 0. \tag{4.148}$$

For the proof of this statement, we refer the reader to (Amann and Hooton, 2013).

While Equation 4.148 is only a necessary and not sufficient condition for Pyragas control to be successful, it is nevertheless very powerful in excluding inefficient control matrices, and serves as a guidance in the design of appropriate control forces. This is in particular true if the orbit to be controlled has an odd number of Floquet multipliers larger than one, since such orbits are notoriously difficult to control. It was even erroneously believed for a long time that Pyragas control cannot be applied to orbits with odd m at all, until an example of successful stabilization was given by Fiedler et al. (2007).

One paradigmatic example for a chaotic system is the Lorenz attractor (Lorenz 1963), which in standard parameters is given by the dynamical equations

$$\dot{x} = 10(y - x),$$
$$\dot{y} = x(28 - z) - y, \tag{4.149}$$
$$\dot{z} = xy - \frac{8}{3}z,$$

where x, y, and z are dynamical variables. The primary unstable orbit ("AB orbit") of the Lorenz system has one Floquet multiplier larger than one, and is therefore difficult to control. Using the guidance from Equation 4.148, Pyragas and Novičenko (2013) found that by using a control matrix of the form

$$K = \kappa \begin{pmatrix} 0 & 0 & 0 \\ -1 & 0 & 0.5 \\ 0 & 0 & 0 \end{pmatrix} \tag{4.150}$$

it is possible to stabilize this orbit. Successful stabilization is shown in Figure 4.25.

5

Complex Networks

The final chapter of this book is devoted to the study of the synchronization properties of systems whose structure of connections forms a complex network. The main focus is to show how the structural properties of the networks affect synchronizability. In addition, we discuss how to assess the stability of the synchronous state in networks, including the case of time-varying structures and multilayer networks. Finally, we describe an experimental setup that can help in studying networked systems.

5.1 Introduction

In the previous chapters we introduced distributed systems and started discussing their dynamics. So far, we have dealt with a rather simple connection topology: each system was either coupled to every other system, or only to its nearest neighbors in some spatial arrangement. However, such simplifications are not always adequate to model real complex systems.

Over the past two decades, a better paradigm – that of networks – has come to be recognized as central for the study of real-world systems. Network models are routinely applied to a wide range of questions in many research areas (Albert and Barabási 2002; Newman 2003; Boccaletti et al. 2006a, 2014). The representation of a complex system as a network, i.e., a set of discrete nodes connected in pairs by discrete links, called *edges*, provides a conceptual simplification that often allows researchers to gain deep analytical insights.

Since the links in a network always connect pairs of nodes, it is possible to represent any particular connection topology as a matrix, which takes the name *adjacency matrix*. In its simplest possible form, an adjacency matrix $\mathbf{A}$ is a 0–1 matrix such that $A_{i,j}$ is 1 if and only if a link exists between node i and node j, and is 0 otherwise.

If no link in a network has a preferential direction, then $A_{i,j} = A_{j,i}$, and $\mathbf{A}$ is a symmetric matrix. Conversely, one may attach a directionality to the links, which is useful when modeling systems in which the interaction between two nodes has univocal origin and destination. In this case, the symmetry condition for the adjacency matrix is not necessarily fulfilled. Finally, if not all the interactions have the same intensity, then the elements of the adjacency matrix are not restricted to the domain {0, 1}. When this happens, the network is said to be *weighted*, and its connectivity matrix is often called a *weighted adjacency matrix*.

To distinguish the two types of matrices without ambiguity, in the following we will use **A** for unweighted, and **W** for weighted adjacency matrices. In an unweighted network, the sum of the elements in the ith row (or column) of **A** is the number k_i of links that the node i forms with the rest of nodes in the network, and is called the *degree*:

$$k_i = \sum_{j=1}^{N} A_{i,j} \ . \tag{5.1}$$

The weighted equivalent of the degree is called the *strength* of a node, and it is defined in a similar fashion:

$$s_i = \sum_{j=1}^{N} W_{i,j} \ . \tag{5.2}$$

Having introduced our basic formalism, we can start considering networks with N nodes representing identical oscillators. Each of the oscillators is assumed to be an m-dimensional, continuous dynamical system. Thus, a vector $\mathbf{x_i}$ with m components is necessary to describe the state of each node.

Since the oscillators are identical, taken independently, the nodes evolve according to the same law. So, in general, one can write that, in the absence of any interaction, the evolution of a node is given by the solution of the system

$$\dot{\mathbf{x_i}} = \mathbf{f}(\mathbf{x_i}) \ . \tag{5.3}$$

In the equation above, **f** is a function that maps $\mathbb{R}^m$ to $\mathbb{R}^m$, and whose particular form defines the local dynamics of the oscillators. Note that here and in the following, to avoid clutter, we omit the explicit time-dependence of the vectors $\mathbf{x_i}$ and of other quantities, except when we want to draw specific attention to it.

To describe how the system evolves when the oscillators are connected in a network, we need to consider not only the local dynamics, described by Equation 5.3, but also how each node is affected by the ones to which it is linked. The law governing the dynamical interaction, which we indicate with **g**, is of course another map from $\mathbb{R}^m$ to $\mathbb{R}^m$, often called the *output function*, and it is in general different from **f**.

We assume that the coupling between the oscillators is diffusive. This means that the effect that node j has on node i is proportional to the difference between $\mathbf{g}(\mathbf{x_j})$

and $\mathbf{g}(\mathbf{x_i})$. Thus, indicating a common interaction strength with σ, and putting ourselves in the most general case of a weighted network, we can write

$$\dot{\mathbf{x_i}} = \mathbf{f}(\mathbf{x_i}) + \sigma \sum_{j=1}^{N} W_{i,j} \left(\mathbf{g}(\mathbf{x_j}) - \mathbf{g}(\mathbf{x_i}) \right)$$

$$= \mathbf{f}(\mathbf{x_i}) + \sigma \left(\sum_{j=1}^{N} W_{i,j} \mathbf{g}(\mathbf{x_j}) - W_{i,j} \mathbf{g}(\mathbf{x_i}) \right)$$

$$= \mathbf{f}(\mathbf{x_i}) + \sigma \left(\sum_{j=1}^{N} W_{i,j} \mathbf{g}(\mathbf{x_j}) - \sum_{j=1}^{N} W_{i,j} \mathbf{g}(\mathbf{x_i}) \right) \tag{5.4}$$

$$= \mathbf{f}(\mathbf{x_i}) + \sigma \left(\sum_{j=1}^{N} W_{i,j} \mathbf{g}(\mathbf{x_j}) - \mathbf{g}(\mathbf{x_i}) \sum_{j=1}^{N} W_{i,j} \right)$$

$$= \mathbf{f}(\mathbf{x_i}) + \sigma \left(\sum_{j=1}^{N} W_{i,j} \mathbf{g}(\mathbf{x_j}) - s_i \mathbf{g}(\mathbf{x_i}) \right) ,$$

where, in the last line, we have used the definition of node strength, Equation 5.2.

From the first line in Equation 5.4, it is clear that when $i = j$ the sum on the right-hand side is 0. Moreover, since the local dynamics are entirely captured by $\mathbf{f}$, the network has no loops, and the diagonal elements of the adjacency matrix all vanish. Thus, from Equation 5.4, it is

$$\dot{\mathbf{x_i}} = \mathbf{f}(\mathbf{x_i}) + \sigma \left(\sum_{j \neq i} W_{i,j} \mathbf{g}(\mathbf{x_j}) - s_i \mathbf{g}(\mathbf{x_i}) \right)$$

$$= \mathbf{f}(\mathbf{x_i}) - \sigma \left(s_i \mathbf{g}(\mathbf{x_i}) - \sum_{j \neq i} W_{i,j} \mathbf{g}(\mathbf{x_j}) \right) \tag{5.5}$$

$$= \mathbf{f}(\mathbf{x_i}) - \sigma \sum_{j=1}^{N} L_{i,j} \mathbf{g}(\mathbf{x_j}) ,$$

where we have defined

$$L_{i,j} = \begin{cases} s_i & \text{if } i = j \\ -W_{i,j} & \text{otherwise.} \end{cases} \tag{5.6}$$

The matrix $\mathbf{L}$ takes the name of *graph Laplacian*, or simply *Laplacian*.

As is clear from its definition, the sum of the elements of each row of the Laplacian vanishes. This means that $\mathbf{L}$ always has a null eigenvalue that corresponds to the eigenvector

$$\frac{1}{\sqrt{N}} (1, 1, \ldots, 1)^{\mathrm{T}} , \qquad (5.7)$$

where the superscript T indicates transposition. In turn, this implies that for any value of σ the network admits a state in which all the oscillators are synchronized, i.e., their trajectories in time are identical, all equal to the same orbit $\mathbf{x_S}(t)$:

$$\mathbf{x_S}(t) \equiv \mathbf{x_1}(t) = \mathbf{x_2}(t) = \cdots = \mathbf{x_N}(t) . \qquad (5.8)$$

This state is an invariant of the dynamics. In fact, if we start the system in a synchronized state, it will remain synchronized. This is most easily seen from the first line of Equation 5.4: if all the state vectors are identical, the evolution reduces to the local dynamics for all the nodes.

As a globally synchronized state on a networked system of this kind always exists and is an invariant of the dynamics, normally one is interested in assessing how stable such a state is. In other words, one wants to estimate how the system in the synchronized state reacts to small perturbations: if the trajectories of the oscillators eventually synchronize again, then the state is said to be stable; if instead the perturbation grows without bound, permanently destroying the synchronization, then the state is unstable.

5.2 Master Stability Function

A powerful, elegant and widely used framework to assess the stability of synchronized states is the Master Stability Function, introduced by Pecora and Carroll (1998).

5.2.1 Derivation of the Master Stability Function

Since the system we are considering is purely deterministic, if the oscillators are set in the synchronized state $\mathbf{x_S}$, they will remain synchronized. However, any perturbation will cause some deviation from $\mathbf{x_S}$. For each oscillator, we can express the effect of such a perturbation as a vector, which we call the *local synchronization error*

$$\delta\mathbf{x_i}(t) = \mathbf{x_i}(t) - \mathbf{x_S}(t) . \qquad (5.9)$$

Note that the local synchronization errors are in general time-dependent. Then, a stable synchronized state is one for which the magnitudes of all the $\delta\mathbf{x_i}$ converge to 0 in the long-time limit. Conversely, an unstable state corresponds to synchronization errors that never vanish. This makes our goal tantamount to study the evolution of the $\delta\mathbf{x_i}$ vectors.

In this treatment, we are assuming that the perturbations are small. Thus, we can approximate the solution by considering a linearized version of the problem, starting from Equation 5.5. To do so, first note that the functions $\mathbf{f}$ and $\mathbf{g}$ are $\mathbb{R}^m \to \mathbb{R}^m$ maps. Thus, the expansion needs to be carried out using a generalization of Taylor's theorem to vector fields. For an infinitely-differentiable function $\mathbf{h}$ between two Euclidean spaces, it is

$$\mathbf{h}(\mathbf{x}) = \sum_{n=0}^{\infty} \frac{1}{n!} \left(J^n \mathbf{h}(\mathbf{x_0}) \right) (\mathbf{x} - \mathbf{x_0})^{\otimes n}, \qquad (5.10)$$

where $J^n \mathbf{h}$ is the 1-contravariant, n-covariant tensor of mixed nth-order partial derivatives of $\mathbf{h}$, the exponent $\otimes n$ indicates nth-order tensor power, and $\mathbf{h}$ is expanded around $\mathbf{x_0}$.

Note that $J^1 \mathbf{h}$ is simply the Jacobian matrix of $\mathbf{h}$. Thus, to linear order, Equation 5.10 becomes

$$\mathbf{h}(\mathbf{x}) \approx \mathbf{h}(\mathbf{x_0}) + J\mathbf{h}(\mathbf{x_0})(\mathbf{x} - \mathbf{x_0}), \qquad (5.11)$$

where J is the Jacobian operator. We can now use Equation 5.11 to linearize Equation 5.5, expanding $\dot{\mathbf{x}}_i$ around the synchronized state to obtain

$$\dot{\mathbf{x}}_i \approx \mathbf{f}(\mathbf{x_S}) + J\mathbf{f}(\mathbf{x_S})(\mathbf{x_i} - \mathbf{x_S}) - \sigma \sum_{j=1}^{N} L_{i,j} \left[\mathbf{g}(\mathbf{x_S}) + J\mathbf{g}(\mathbf{x_S})(\mathbf{x_j} - \mathbf{x_S}) \right]$$

$$= \left(\mathbf{f}(\mathbf{x_S}) - \sigma \sum_{j=1}^{N} L_{i,j} \mathbf{g}(\mathbf{x_S}) \right) + J\mathbf{f}(\mathbf{x_S})(\mathbf{x_i} - \mathbf{x_S})$$

$$- \sigma \sum_{j=1}^{N} L_{i,j} J\mathbf{g}(\mathbf{x_S})(\mathbf{x_j} - \mathbf{x_S})$$

$$= \dot{\mathbf{x}}_S + J\mathbf{f}(\mathbf{x_S})(\mathbf{x_i} - \mathbf{x_S}) - \sigma \sum_{j=1}^{N} L_{i,j} J\mathbf{g}(\mathbf{x_S})(\mathbf{x_j} - \mathbf{x_S}),$$

$$(5.12)$$

hence we get

$$\dot{\mathbf{x}}_i - \dot{\mathbf{x}}_S \approx J\mathbf{f}(\mathbf{x_S}) \delta\mathbf{x}_i - \sigma \sum_{j=1}^{N} L_{i,j} J\mathbf{g}(\mathbf{x_S}) \delta\mathbf{x}_j, \qquad (5.13)$$

where we used the definition of local synchronization error, Equation 5.9. Defining now $\delta\dot{\mathbf{x}}_i \equiv \dot{\mathbf{x}}_i - \dot{\mathbf{x}}_S$, and noting that the Jacobian of $\mathbf{g}$ computed on the synchronized state does not depend on j, we finally get

$$\delta\dot{\mathbf{x}}_i \approx J\mathbf{f}(\mathbf{x_S}) \delta\mathbf{x}_i - \sigma J\mathbf{g}(\mathbf{x_S}) \sum_{j=1}^{N} L_{i,j} \delta\mathbf{x}_j. \qquad (5.14)$$

The equation we just derived approximates the evolution of each local synchronization error to linear order.

It is now convenient to pass to a global formalism. To do so, we first introduce the Nm-dimensional state vector $\mathbf{X} \equiv (\mathbf{x_1}^T, \mathbf{x_2}^T, \ldots, \mathbf{x_N}^T)^T$. Similarly, we can define the *global synchronization error* vector as $\delta\mathbf{X} \equiv (\delta\mathbf{x_1}^T, \delta\mathbf{x_2}^T, \ldots, \delta\mathbf{x_N}^T)^T$ and its derivative as $\delta\dot{\mathbf{X}} \equiv (\delta\dot{\mathbf{x}}_1^T, \delta\dot{\mathbf{x}}_2^T, \ldots, \delta\dot{\mathbf{x}}_N^T)^T$. Now note that the first term on the right-hand side of Equation 5.14 only mixes the ith state vector with itself, while the second term mixes it with a combination of all other state vectors whose coefficients are given by the Laplacian elements $L_{i,j}$. This suggests a way to easily write a linearized evolution equation directly for the global synchronization error.

Since $\delta\dot{\mathbf{X}}$ has Nm elements, we can write it as the product of an Nm-dimensional square matrix and $\delta\mathbf{X}$. We decompose the matrix itself into the sum of two parts. Indicating the N-dimensional identity matrix as $\mathbb{1}_N$, the first part is just $\mathbb{1}_N \otimes J\mathbf{f}(\mathbf{x_S})$. This is a block-diagonal matrix with N square blocks of dimension m, each of which is the Jacobian of $\mathbf{f}$ computed on the synchronized state, accounting for the self-mixing of the state vectors in Equation 5.14. The second part of the matrix is $-\sigma\mathbf{L} \otimes J\mathbf{g}(\mathbf{x_S})$. This is itself a matrix constituted by N^2 square blocks of dimension m; each block of coordinates (i, j) consists of the scalar $-\sigma L_{i,j}$ multiplied by the Jacobian of $\mathbf{g}$ computed on the synchronized state. Then, one can finally write

$$\delta\dot{\mathbf{X}} \approx (\mathbb{1}_N \otimes J\mathbf{f}(\mathbf{x_S}) - \sigma\mathbf{L} \otimes J\mathbf{g}(\mathbf{x_S}))\, \delta\mathbf{X}\,. \tag{5.15}$$

The equation above directly describes the evolution of the global synchronization error, formalized as a vector of vectors. However, due to the $\mathbf{L} \otimes J\mathbf{g}$ term, it still mixes different components.

Our aim is to rewrite Equation 5.15 so that the whole part in parentheses is a block-diagonal matrix. To achieve this, we diagonalize $\mathbf{L}$. Note that this is always possible, because $\mathbf{L}$ is a real symmetric matrix, and therefore it can always be diagonalized by a basis of $\mathbb{R}^N$. Then, let $\mathbf{V}$ be the matrix whose columns are the eigenvectors of $\mathbf{L}$, so that $\mathbf{V}^T\mathbf{L}\mathbf{V} = \mathbf{D}$, where $\mathbf{D}$ is the diagonal matrix of the eigenvalues of $\mathbf{L}$. Multiplying Equation 5.15 on the left by $(\mathbf{V}^T \otimes \mathbb{1}_m)$ we get

$$\begin{aligned}
(\mathbf{V}^T \otimes \mathbb{1}_m)\, \delta\dot{\mathbf{X}} &\approx (\mathbf{V}^T \otimes \mathbb{1}_m)\, (\mathbb{1}_N \otimes J\mathbf{f}(\mathbf{x_S}) - \sigma\mathbf{L} \otimes J\mathbf{g}(\mathbf{x_S}))\, \delta\mathbf{X} \\
&= \big[(\mathbf{V}^T \otimes \mathbb{1}_m)\, (\mathbb{1}_N \otimes J\mathbf{f}(\mathbf{x_S})) \\
&\quad - (\mathbf{V}^T \otimes \mathbb{1}_m)\, (\sigma\mathbf{L} \otimes J\mathbf{g}(\mathbf{x_S})) \big]\, \delta\mathbf{X}\,.
\end{aligned} \tag{5.16}$$

Now, use the relation

$$(\mathbf{M_1} \otimes \mathbf{M_2})\, (\mathbf{M_3} \otimes \mathbf{M_4}) = (\mathbf{M_1}\mathbf{M_3}) \otimes (\mathbf{M_2}\mathbf{M_4}) \tag{5.17}$$

to obtain

$$\left(\mathbf{V}^{\mathrm{T}} \otimes \mathbb{1}_m\right) \delta \dot{\mathbf{X}} \approx \left[\mathbf{V}^{\mathrm{T}} \otimes J\mathbf{f}\left(\mathbf{x}_{\mathbf{S}}\right) - \left(\sigma \mathbf{V}^{\mathrm{T}} \mathbf{L}\right) \otimes J\mathbf{g}\left(\mathbf{x}_{\mathbf{S}}\right)\right] \delta \mathbf{X}$$
$$= \left[\mathbf{V}^{\mathrm{T}} \otimes J\mathbf{f}\left(\mathbf{x}_{\mathbf{S}}\right) - \left(\sigma \mathbf{D}\mathbf{V}^{\mathrm{T}}\right) \otimes J\mathbf{g}\left(\mathbf{x}_{\mathbf{S}}\right)\right] \delta \mathbf{X} . \tag{5.18}$$

In this last expression, $\mathbf{V}^{\mathrm{T}}$ is N-dimensional, while $J\mathbf{f}$ and $J\mathbf{g}$ are m-dimensional. Multiplying each by the identity matrix of the correct dimension, we get

$$\left(\mathbf{V}^{\mathrm{T}} \otimes \mathbb{1}_m\right) \delta \dot{\mathbf{X}} \approx \left[\left(\mathbb{1}_N \mathbf{V}^{\mathrm{T}}\right) \otimes \left(J\mathbf{f}\left(\mathbf{x}_{\mathbf{S}}\right) \mathbb{1}_m\right)\right.$$
$$\left. - \left(\sigma \mathbf{D}\mathbf{V}^{\mathrm{T}}\right) \otimes \left(J\mathbf{g}\left(\mathbf{x}_{\mathbf{S}}\right) \mathbb{1}_m\right)\right] \delta \mathbf{X} . \tag{5.19}$$

Using again Equation 5.17, it is

$$\left(\mathbf{V}^{\mathrm{T}} \otimes \mathbb{1}_m\right) \delta \dot{\mathbf{X}} \approx \left[\left(\mathbb{1}_N \otimes J\mathbf{f}\left(\mathbf{x}_{\mathbf{S}}\right)\right) \left(\mathbf{V}^{\mathrm{T}} \otimes \mathbb{1}_m\right)\right.$$
$$\left. - \left(\sigma \mathbf{D} \otimes J\mathbf{g}\left(\mathbf{x}_{\mathbf{S}}\right)\right) \left(\mathbf{V}^{\mathrm{T}} \otimes \mathbb{1}_m\right)\right] \delta \mathbf{X} \tag{5.20}$$
$$= \left(\mathbb{1}_N \otimes J\mathbf{f}\left(\mathbf{x}_{\mathbf{S}}\right) - \sigma \mathbf{D} \otimes J\mathbf{g}\left(\mathbf{x}_{\mathbf{S}}\right)\right) \left(\mathbf{V}^{\mathrm{T}} \otimes \mathbb{1}_m\right) \delta \mathbf{X} .$$

Letting $\boldsymbol{\eta} \equiv \left(\mathbf{V}^{\mathrm{T}} \otimes \mathbb{1}_m\right) \delta \mathbf{X}$, we finally get

$$\dot{\boldsymbol{\eta}} \approx \left(\mathbb{1}_N \otimes J\mathbf{f}\left(\mathbf{x}_{\mathbf{S}}\right) - \sigma \mathbf{D} \otimes J\mathbf{g}\left(\mathbf{x}_{\mathbf{S}}\right)\right) \boldsymbol{\eta} . \tag{5.21}$$

The equation above governs the behavior of the vector-of-vectors $\boldsymbol{\eta}$ without mixing its components. The treatment effectively decomposes the time-evolution of the global synchronization error into decoupled eigenmodes identified by the eigenvectors of $\mathbf{L}$. This is perhaps more evident when rewriting Equation 5.21 as a system of N equations

$$\dot{\boldsymbol{\eta}}_{\mathbf{j}} \approx \left(J\mathbf{f}\left(\mathbf{x}_{\mathbf{S}}\right) - \sigma \lambda_j J\mathbf{g}\left(\mathbf{x}_{\mathbf{S}}\right)\right) \boldsymbol{\eta}_{\mathbf{j}} . \tag{5.22}$$

The first eigenvalue of $\mathbf{L}$, λ_1, is always 0, as shown before. Thus, the equation for $\boldsymbol{\eta}_1$ is simply the variational equation

$$\dot{\boldsymbol{\eta}}_1 \approx J\mathbf{f}\left(\mathbf{x}_{\mathbf{S}}\right) \boldsymbol{\eta}_1 . \tag{5.23}$$

The other eigenmodes, $\boldsymbol{\eta}_2, \boldsymbol{\eta}_3, \ldots, \boldsymbol{\eta}_{\mathbf{N}}$, are separated from $\boldsymbol{\eta}_1$. Therefore, a necessary condition for stability is that they all vanish in the long-time limit.

Notice that Equation 5.8 defines a subset S of the phase space of the system, on which the evolution occurs when the oscillators synchronize. This is because each of the $N-1$ equalities in Equation 5.8 imposes m constraints, one per dimension of the oscillators. Each point of S corresponds to some given position and velocity of the synchronized oscillators. Thus, S is an m-dimensional manifold, embedded in the $2Nm$-dimensional phase space of the system. In the case of identical synchronization, which we consider here, S is actually a hyperplane, whose orientation is determined by the equalities of Equation 5.8.

To estimate the stability of S, first note that Equation 5.23 only depends on the local dynamics of the oscillators. Thus, $\boldsymbol{\eta}_1$ measures the perturbations to the motion along the synchronization manifold $\mathcal{S}$. Also, we have seen from Equation 5.21 that the other components of $\boldsymbol{\eta}$ are transverse to $\boldsymbol{\eta}_1$, and therefore to $\mathcal{S}$.

After a transient, the evolution of $\boldsymbol{\eta}$ is such that $|\boldsymbol{\eta}|\,(t) \sim e^{\Lambda_{max}t}$, where Λ_{max} is the largest Lyapunov exponent associated with the system in Equation 5.22. Note that the norm of $\boldsymbol{\eta}$ is to be computed over all its components *except* $\boldsymbol{\eta}_1$, since we are interested in the eigenmodes transverse to the synchronization manifold. The expression of Λ_{max} as a function of a generic parameter ν takes the name of *Master Stability Function* (MSF). Letting $\nu_j \equiv \sigma\lambda_j$, the MSF allows one to estimate the stability of each eigenmode with eigenvalue λ_j by simply considering the sign of $\Lambda_{max}\left(\nu_j\right)$.

5.2.2 Classes of Synchronizability

For the synchronization manifold to be transversally stable, one needs $\Lambda_{max}\left(\nu_j\right)$ to be negative for all the $\nu_j \geqslant 0$.

Notice that $\Lambda_{max}\,(0)$ is always greater than or equal to 0, because it corresponds to the maximum Lyapunov exponent of a single isolated (not networked) dynamical system. In fact, an identically vanishing ν_j implies that either $\sigma = 0$, or all $\lambda_j = 0$. In the former case, there is no coupling between neighbors on the network; in the latter, $\mathbf{L} = \mathbf{0}$ and the network has no links. Either way, the systems are effectively isolated, and the dynamics are given by the eigenmode associated with the synchronization manifold. Then, the evolution of the system reduces to

$$\dot{\mathbf{x}}_i = \mathbf{f}\left(\mathbf{x}_i\right) \tag{5.24}$$

for all i, and Λ_{max} is just the maximum Lyapunov exponent of a single system, which is positive if the local dynamics are chaotic, or zero if it supports a periodic orbit.

Then, there are three possible behaviors of the MSF: (Boccaletti et al. 2006a)

(I) $\Lambda_{max}\left(\nu_j\right)$ is a monotonically increasing function, always nonnegative, and never crossing the ν_j axis;

(II) $\Lambda_{max}\left(\nu_j\right)$ is a monotonically decreasing function that crosses the ν_j axis at $\nu = \nu_C \geqslant 0$, becoming negative for all $\nu_j > \nu_C$;

(III) $\Lambda_{max}\left(\nu_j\right)$ is a V-shaped function that crosses the ν_j axis in two points ν_{min} and ν_{max}, between which it is negative.

This is an important result, because it holds for all possible choices of local and output functions $\mathbf{f}$ and $\mathbf{g}$. The three cases, schematically illustrated in Figure 5.1, correspond to different classes of synchronizability of the network.

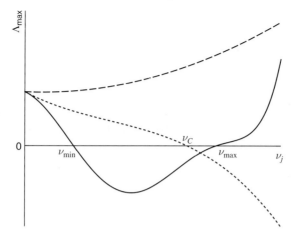

Figure 5.1 Sketch of the possible behaviors of the MSF. The dashed line corresponds to systems in class I, whose synchronized states are never stable. The dotted line corresponds to systems in class II, whose synchronized states are stable for any coupling stronger than a critical value. The solid line corresponds to systems in class III, which need two conditions to be satisfied for stability.

Synchronization in systems of class I is trivially never stable. In fact, no matter the value of the coupling σ, the MSF for these systems is always positive, which indicates that a nonzero synchronization error never vanishes.

Systems in class II are an almost opposite case, since for them Λ_{max} is negative for all values of ν_j greater than a critical value ν_C. Then, all is needed to obtain stable synchronization is to guarantee that $\nu_C < \sigma\lambda_2 = \nu_2$, by choosing a coupling strength $\sigma > \nu_C/\lambda_2$. Since all other eigenvalues are greater than or equal to λ_2, this implies that $\nu_j > \nu_C$ for all j. Notice that in systems belonging to this class, the synchronization threshold (the minimal value of σ for which synchronization becomes stable) is inversely proportional to the second smallest eigenvalue λ_2. In other words, the larger is λ_2, the smaller is the synchronization threshold, and one can use this property to construct a network that optimizes synchronization, once the functions $\mathbf{f}$ and $\mathbf{g}$ are given.

Class III is possibly the most intriguing, and indeed many choices for the functions $\mathbf{f}$ and $\mathbf{g}$ cause the dynamics to be in this class (Barahona and Pecora 2002). To ensure that all ν_j lie within the interval (ν_{min}, ν_{max}), in these systems one needs to satisfy two conditions at the same time:

$$\nu_{min} < \sigma\lambda_2 \tag{5.25}$$
$$\sigma\lambda_N < \nu_{max} . \tag{5.26}$$

From Equation 5.25 it follows that it has to be

$$\frac{\nu_{min}}{\lambda_2} < \sigma . \tag{5.27}$$

Substituting in Equation 5.26, it is

$$v_{min} \frac{\lambda_N}{\lambda_2} < v_{max} , \tag{5.28}$$

hence

$$\frac{\lambda_N}{\lambda_2} < \frac{v_{max}}{v_{min}} . \tag{5.29}$$

Equivalently, from Equation 5.26 one gets

$$\sigma < \frac{v_{max}}{\lambda_N} , \tag{5.30}$$

from which it is

$$v_{min} < v_{max} \frac{\lambda_2}{\lambda_N} , \tag{5.31}$$

and Equation 5.29 follows. It is easy to see why Equation 5.29 is often studied as the main necessary condition for the stability of synchronization in systems of class III: only if it holds, values of the coupling strength σ between v_{min}/λ_2 and v_{max}/λ_N cause all the v_j to lie within the stability interval.

Equation 5.29 also provides a measure of the synchronizability of a networked system in class III via the eigenratio

$$r = \frac{\lambda_N}{\lambda_2} . \tag{5.32}$$

The smaller is r, the easier it is to satisfy the condition in Equation 5.29. Therefore, one can say that, given $\mathbf{f}$ and $\mathbf{g}$, the networks that minimize r are those where synchronization is most likely to be stable. The criterion of minimizing r has been largely used as a guide to design networks where synchronization properties are enhanced. Since $r \geqslant 1$, the best networks are those for which the entire nonnull spectrum of the Laplacian is degenerate, resulting in $r = 1$. Such structures are often called *maximally synchronizable networks*.

Note that the behavior of the MSF only depends on the functions $\mathbf{f}$ and $\mathbf{g}$, as it is clear from Equation 5.22. Thus, the synchronizability class of a system depends exclusively on its internal dynamics and on the output function through which individual nodes interact with the rest of the network.

A prototypical example is given by a system of coupled Rössler oscillators (Rössler 1976). Each oscillator is a three-dimensional dynamical system whose local dynamics are

$$\mathbf{f}(\mathbf{x_i}) = \begin{pmatrix} -x_{i_2} - x_{i_3} \\ x_{i_1} + a x_{i_2} \\ b + x_{i_3} \left(x_{i_1} - c \right) \end{pmatrix} , \tag{5.33}$$

Different choices of the scalar parameters a, b, and c correspond to different types of dynamics. For example, setting $a = 0.2$, $b = 0.2$, and $c = 9$ ensures that the dynamics are chaotic. With these values of the parameters, one can choose the synchronization class of the global dynamics by selecting one of three simple choices for $\mathbf{g}$ (Boccaletti et al. 2006a). Specifically, choosing

$$\mathbf{g}\,(\mathbf{x_i}) = \begin{pmatrix} 0 \\ 0 \\ x_{i_3} \end{pmatrix} \tag{5.34}$$

causes the system to be in class I, and synchronized states to be always unstable. Choosing instead

$$\mathbf{g}\,(\mathbf{x_i}) = \begin{pmatrix} 0 \\ x_{i_2} \\ 0 \end{pmatrix} \tag{5.35}$$

yields a system of class II with $\nu_C = 0.1445$. Finally,

$$\mathbf{g}\,(\mathbf{x_i}) = \begin{pmatrix} x_{i_1} \\ 0 \\ 0 \end{pmatrix} \tag{5.36}$$

selects class III, with $\nu_{\min} = 0.181$ and $\nu_{\max} = 4.615$.

The beauty and elegance of the MSF approach is that it successfully disentangles the dynamics, which determine the class of a system, from the network topology, which provides the values of the Laplacian eigenvalues and ultimately determines whether the dynamical units can be synchronized with a particular structure of connections. Thus, in the following we briefly discuss the spectral properties of several common types of complex networks, as they directly influence their ability to synchronize.

5.3 Small-World Networks

One of the first network models created with the aim of reproducing some of the features found in real-world systems was introduced by the Australian researcher Duncan Watts and his doctoral advisor, American scientist Steven Strogatz (1998), who used a ring-rewiring method to yield so-called *small-world networks* (SW). This is perhaps the simplest example of a complex network, and thus it is our starting point to discuss network synchronizability. The rewiring procedure, which we describe more in detail in the following sections, introduces nonlocal shortcuts in a ring lattice, thus connecting distant nodes. As we will see, the number of necessary shortcuts to create a SW network is effectively quite small. So, to understand the synchronization properties of such networks, it is useful to study those of the initial ring lattices.

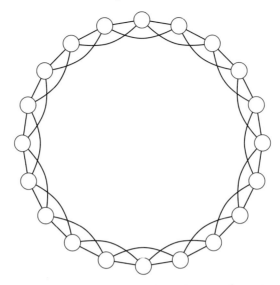

Figure 5.2 Example of a ring lattice. This starting configuration has $N = 20$ nodes. Each is connected to its $m = 2$ nearest neighbors in either direction, so that all nodes have degree $k = 4$.

5.3.1 Ring Lattices

The starting configuration in the creation of a small-world network is that of a ring of N nodes, each connected to m neighbors in each direction (see Figure 5.2). Therefore, the elements of its Laplacian $\mathbf{L}^0$ are

$$L_{i,j}^0 = \begin{cases} 2m & \forall\, i = j \\ -1 & \forall\, 0 < |i - j| \leqslant m \\ 0 & \text{otherwise .} \end{cases} \tag{5.37}$$

An easy way of finding the smallest nonzero eigenvalue and the largest is to notice that $\mathbf{L}^0$ is a circulant matrix defined by the vector

$$\mathbf{V} = \left(2m, \{-1\}^m, \{0\}^{N-2m-1}, \{-1\}^m \right)^{\mathrm{T}} . \tag{5.38}$$

Thus, the eigenvalues of $\mathbf{L}^0$ are identified by the discrete Fourier transform coefficients of $\mathbf{V}$ (Monasson 1999):

$$\lambda_n = 2m - 2 \sum_{t=1}^{m} \cos\left(\frac{2\pi\,(n-1)}{N} t \right) . \tag{5.39}$$

Also, the components of the corresponding eigenvectors $\mathbf{v_n}$ are

$$\mathbf{v}_{nj} = \frac{1}{\sqrt{N}} e^{\frac{1}{N} 2\pi \mathrm{i}(n-1)j} . \tag{5.40}$$

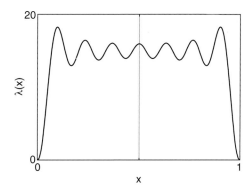

Figure 5.3 Properties of $\lambda(x)$ (Equation 5.41). The function, plotted for $m = 7$, has its global minimum at 0 and it is even with respect to the axis $x = {}^1/_2$.

Computing the sum, Equation 5.39 can be rewritten as

$$\lambda(x) = 2m + 1 - \csc(\pi x)\sin(\pi x(2m+1)) , \qquad (5.41)$$

where we put

$$x = \frac{n-1}{N} \qquad (5.42)$$

for the sake of simplicity.

Equation 5.41 defines a continuous function that provides the eigenvalues of $\mathbf{L_0}$ at all x corresponding to integer values of n. Note that the function is periodic with period N, it vanishes at all integer values of x, and it satisfies $\lambda(1/2 + x) = \lambda(1/2 - x)$. Also, within the interval of interest $[0, 1)$, its global minimum is at 0 (see Figure 5.3). This allows one to estimate the smallest nonzero eigenvalue of $\mathbf{L_0}$, also known as the spectral gap, which is $\lambda_2 = \lambda(1/N)$. Since for large N it is $1/N \ll 1$, one can expand Equation 5.41 around $x = 0$ to obtain

$$\lambda\left(\frac{1}{N}\right) = \frac{2}{3}\pi^2 m(m+1)(2m+1)x^2 + O\left(x^4\right) , \qquad (5.43)$$

which gives

$$\lambda_2 = \frac{2\pi^2 m(m+1)(2m+1)}{3} \frac{}{N^2} + O\left(N^{-4}\right) . \qquad (5.44)$$

The next step is finding the largest eigenvalue λ_N. Asymptotically, the density of eigenvalues given by $\lambda(x)$ in the $[0, 1)$ interval is arbitrarily large. Thus, any point is arbitrarily close to an eigenvalue, and the largest eigenvalue λ_N is well approximated by the global maximum of $\lambda(x)$. The total number of maxima of $\lambda(x)$ is exactly m (see also Figure 5.3). To see this, notice that there is always a range of values of x for which the sine in Equation 5.41 is negative, due to the $(2m + 1)$ factor within its argument. At the same time, the cosecant is never negative, since its argument is constrained within $[0, \pi)$. As in this range the cosecant–sine product has a negative sign, the local maxima of λ correspond to all values of x for which the argument of the sine is $3\pi/2$ modulo 2π:

$$(2m + 1)\,\pi x = \frac{3}{2}\pi + 2z\pi\,, \quad z \in \{0, 1, \ldots, m - 1\}\,, \tag{5.45}$$

hence

$$x = \frac{3}{2\,(2m + 1)} + \frac{2z}{2m + 1}\,. \tag{5.46}$$

For all these values of x, the sine is always -1. Then, the global maximum is at the x for which the cosecant is largest. Since λ is even with respect to $x = 1/2$, we can limit our search to the interval $\left[0, 1/2\right]$. In this interval, the cosecant is monotonically decreasing. Thus, its largest value corresponds to the smallest candidate argument, that is, to $z = 0$. This yields directly

$$\lambda_N \approx 2m + 1 + \csc\left(\frac{3\pi}{2\,(2m + 1)}\right)\,. \tag{5.47}$$

For large m, this expression can be further approximated to

$$\lambda_N \approx 2m + 1 + \frac{2\,(2m + 1)}{3\pi} = (2m + 1)\left(1 + \frac{2}{3\pi}\right)\,. \tag{5.48}$$

This shows that, for large ring lattices with a sufficiently large number of local connections, the eigenvalue ratio

$$\frac{\lambda_N}{\lambda_2} \approx \left(\frac{3}{2} + \frac{1}{\pi}\right)\frac{N^2}{\pi^2 m\,(m + 1)} = \frac{(3\pi + 2)\,N^2}{2\pi^3 m\,(m + 1)} \sim \frac{N^2}{m^2}\,. \tag{5.49}$$

Since the neighbors in the starting configuration are by definition local, it is always $m \ll N$. This means that, for Equation 5.29 to be satisfied in a specific system of oscillators on a ring lattice, it has to be

$$m \gtrsim N\sqrt{\frac{\nu_{\min}}{\nu_{\max}}}\,. \tag{5.50}$$

With the locality of the neighborhoods, this equation implies that ring lattices are quite hard to synchronize, and their synchronizability depends mostly on the ratio of the stability interval bounds, which is a quantity determined only by the properties of the node dynamics.

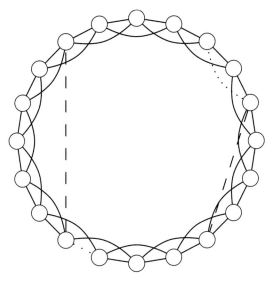

Figure 5.4 Example of a small-world network. The starting configuration was that of Figure 5.2, with 20 nodes, each linked to the two nearest neighbors in either direction. The rewiring probability was $p = 0.05$, and resulted in two links (dotted lines) being erased and replaced with shortcuts (dashed lines).

5.3.2 The Watts–Strogatz Model

Watts and Strogatz (1998) introduced a very successful procedure to create a type of structure known as small-world networks. Starting from a ring lattice, the model consists of the following steps: (i) proceed around the ring always in the same direction (clockwise or counterclockwise), and (ii) systematically consider each edge, and rewire it to a randomly chosen node with probability p, avoiding the creation of multiple edges (Figure 5.4). The procedure is complete when all edges have been considered once.

The resulting networks are characterized by two main features. First, the average length of the shortest path between any two nodes $\langle d\,(i, j)\rangle$ increases only logarithmically with the number of nodes N. While this is similar to what happens in Erdős–Rényi random graphs, it is in sharp contrast with the general behavior of d-dimensional lattices, for which the shortest path grows like $\sqrt[d]{N}$. Also, for a given starting configuration, $\langle d\,(i, j)\rangle$ quickly decreases with increasing rewiring probability.

Second, for a wide range of p, the rewiring procedure introduces no significant change in the average local clustering coefficient c_i, defined for each node i as

$$c_i = \frac{1}{k_i\,(k_i - 1)} \sum_{j=1}^{N} A_{i,j} \sum_{r=1}^{N} A_{i,r} A_{j,r} . \qquad (5.51)$$

The coefficient c_i measures the fraction of mutual neighbors among the common neighbors of node i. Thus, it is straightforward to compute it for the starting configuration: given a node i, its m nearest neighbors in one direction are fully connected; additionally, each of the remaining m neighbors is connected to $m - 1$ other neighbors of i. Thus, the starting local clustering coefficient is

$$c_0 = \frac{2}{2m\,(2m-1)} \left[\frac{m\,(m-1)}{2} + m\,(m-1) \right]$$

$$= \frac{2}{2m\,(2m-1)} \frac{3}{2} m\,(m-1) \qquad\qquad (5.52)$$

$$= \frac{3m-3}{4m-2}.$$

Equivalently, since $k = 2m$, it is

$$c_0 = \frac{3\frac{k}{2} - 3}{1\frac{k}{2} - 2} = \frac{3k-6}{4k-4} = \frac{3\,(k-2)}{4\,(k-1)}. \qquad\qquad (5.53)$$

Either way, for sufficiently connected ring lattices, the clustering coefficient converges to $3/4$, and it remains high in small-world networks (see Figure 5.5).

This shows that SW networks have two important properties of real-world systems, namely the short average distance between nodes, and the high local clustering. The model we briefly described obtains these results by introducing links that act as shortcuts between distant nodes. One could then intuitively

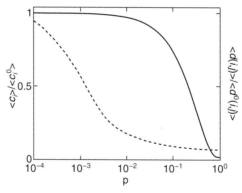

Figure 5.5 The small-world regime. For a wide range of rewiring probabilities p, the average local clustering coefficient of a SW network does not deviate substantially from that of the initial ring lattice used for its construction (solid line). At the same time, the average distance between pairs of nodes decreases quickly with p (dotted line). The data shown are averages over 100 SW realizations created from a starting ring of 1000 nodes with $m = 5$, and error bars are smaller than the thickness of the lines. Notice the logarithmic scale for p.

assume that, promoting the exchange of information over longer distances, the SW property favors synchronization. To test this assumption, we estimate λ_2 for SW networks.

From Figure 5.5 it is evident that the small-world range, in which the path length is small, but the clustering has not yet decreased, extends over very low values of p. Thus, the number of shortcuts to introduce in order to create a SW network is quite small when compared to the total number of links. This suggests that we can treat the SW Laplacian $\mathbf{L}^S$ perturbatively, starting from the Laplacian $\mathbf{L}^0$ of a ring lattice:

$$\mathbf{L}^S = \mathbf{L}^0 + \mathbf{L}^1 , \tag{5.54}$$

where $\mathbf{L}^1$ is the Laplacian of the graph of shortcuts. If each node in the ring lattice has m neighbors per direction, the total number of links available for rewiring is mN. Similarly, the number of possible shortcuts is $(N^2 - N - 2mN)/2$. Therefore, if each link is rewired with probability p, the probability of creating each available shortcut is

$$p' = \frac{2pm}{N - 2m - 1} . \tag{5.55}$$

Thus, on average, $\mathbf{L}^1$ is a matrix in which each "free" element has probability p' of being -1, and probability $1 - p'$ of being 0. The effect of this perturbation on the spectrum of $\mathbf{L}^0$ can be described in terms of the resolvent

$$\mathbf{R}(z) = (z\mathbb{1} - \mathbf{L})^{-1} , \quad z \in \mathbb{C} , \tag{5.56}$$

which determines the spectral properties of any graph Laplacian, or indeed any operator. In particular, the small-world resolvent $\mathbf{R}^S$ obeys Dyson's equation

$$\mathbf{R}^S = \mathbf{R}^0 + \mathbf{R}^0 \mathbf{L}^1 \mathbf{R}^S . \tag{5.57}$$

In this expression, $\mathbf{L}^1$ can be interpreted as a self-energy operator Σ, whose contribution determines the expected spectral shift on Equation 5.41 due to the introduced perturbation. Stopping the canonical expansion to first order, it is

$$\Sigma(x) = \mathbf{v_n}^\dagger \mathbf{L}^1 \mathbf{v_n} , \tag{5.58}$$

where x and n are related by Equation 5.42, the vectors $\mathbf{v}$ are given by Equation 5.40, and the equation holds for values of x corresponding to integer n. Evaluating the expression yields an expected shift of $2pm$. As this increases λ_2, the result suggests that synchronizability in small-world networks is enhanced with respect to ring lattices in a way proportional to the number of shortcuts placed, as one would intuitively expect.

5.4 Preferential Attachment Networks

Small-world networks reproduce two important features found in real-world systems, namely high clustering and low distance between nodes. However, they still do not provide a very realistic model. One main reason is that in small-world networks the distribution of the node degrees is centred around $2m$, whereas in many real-world systems it is highly heterogeneous.

In fact, numerous studies have shown that the degrees in real-world networks are most often distributed according to a power law:

$$P(k) = \frac{k^{-\gamma}}{H_{N-1,\gamma}}. \tag{5.59}$$

In the equation above, the normalization constant

$$H_{N-1,\gamma} = \sum_{k-1}^{N-1} k^{-\gamma} \tag{5.60}$$

is the $(N-1)$th generalized harmonic number of exponent γ.

Such structures take the name of scale-free networks (SF), since power laws are the only functional relations that do not change at different scales. Scale-free networks are ubiquitous, having been found in natural and engineered systems of the most diverse nature, and have generated considerable research interest since the beginning of the 2000s. Several generative models and techniques have been proposed for this type of network, including both static and dynamic processes. One of the most widely used is the so-called *preferential attachment* (PA) model introduced by Barabási and Albert (1999).

To create a PA network in its most typical form, one starts with an initial seed, given by a small group of all-to-all connected nodes. At each step, a new node is added to the network, and linked to an already present node chosen randomly with a probability proportional to its degree. The process continues until the desired network size is reached, yielding a network whose power-law exponent γ is 3.

The value of the exponent can be tuned by altering the probability of choosing target nodes for linking (Krapivsky et al. 2000): to obtain a chosen γ, leave the selection probability of degree-1 nodes proportional to 1, and make that of degree-k nodes proportional to μk, where

$$\mu = \frac{8}{4\gamma^2 - 12\gamma + 8}. \tag{5.61}$$

Note that this equation implies that the range of possible exponents is $(2, \infty)$. The preferential attachment procedure can be easily extended to the case where each new node forms $m > 1$ connections with already existing nodes, provided that the initial seed is constituted by at least m nodes. Here, however, we only consider the case of a single link added per step.

It is to be stressed that PA networks are only a subset of SF networks. In fact, while PA networks inherit many statistical properties of SF ones, including the exponent limits (del Genio et al. 2011), the two sets have important structural differences. The most relevant one in our context is that the construction procedure outlined above always yields connected networks, while SF networks can in general consist of more than one connected component. This is straightforward to verify using other construction methods, such as configuration model algorithms (Newman et al. 2001), or degree-based graph sampling approaches (del Genio et al. 2010; Kim et al. 2011; Bassler et al. 2015). However, since one cannot discuss synchronization if a network is not connected, in the following we limit ourselves to PA networks.

To estimate the eigenratio, Equation 5.29, for PA networks, first note that, according to Kim and collaborators (2007), the spectral density of the Laplacian of PA networks in the thermodynamic limit has a nonzero gap and a power-law tail with the same exponent as the degree distribution:

$$\lambda(x) \propto x^{-\gamma} . \tag{5.62}$$

Then, one can compute the expected maximum eigenvalue up to a multiplicative constant by solving for λ_N the equation

$$N \int_{\lambda_N}^{N} x^{-\gamma} dx = 1. \tag{5.63}$$

This yields

$$\frac{N}{1-\gamma} \left(N^{1-\gamma} - \lambda_N^{1-\gamma} \right) = 1 , \tag{5.64}$$

which, for very large N, becomes

$$\frac{N}{\gamma - 1} \lambda_N^{1-\gamma} = 1 , \tag{5.65}$$

hence

$$\lambda_N^{1-\gamma} \propto \frac{\gamma - 1}{N} \tag{5.66}$$

and

$$\lambda_N \propto \left(\frac{\gamma - 1}{N} \right)^{\frac{1}{1-\gamma}} \sim N^{\frac{1}{\gamma-1}} . \tag{5.67}$$

This estimate is confirmed by numerical simulations, which also show that $\lambda_2 \sim 1/N$ for a range of exponents and, as a consequence, the eigenratio scales as

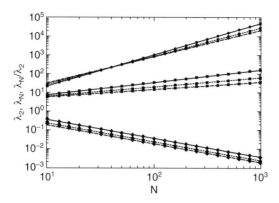

Figure 5.6 Extreme Laplacian eigenvalues and eigenratio for preferential attachment networks. The smallest nonzero eigenvalue λ_2 (diamonds) scales like $^1/_N$, the largest eigenvalue λ_N (squares) scales like $N^{1/\gamma-1}$, and the eigenratio $^{\lambda_N}/_{\lambda_2}$ (circles) scales like $N^{\gamma/\gamma-1}$. The data shown are averages over 1000 realizations of PA networks for $\gamma = 2.5$ (solid lines), $\gamma = 3$ (dotted lines) and $\gamma = 3.5$ (dashed lines).

$N^{\gamma/\gamma-1}$ (Figure 5.6). Since $\gamma/(\gamma-1)$ is always greater than 2 for $\gamma > 2$, PA networks of class-III systems are always more synchronizable than ring lattices and SW networks, whose eigenratio (Equation 5.49) scales like N^2.

Nonetheless, it is to be noticed that, for all models considered, λ_2 vanishes in the limit of an infinite network, making the networks not synchronizable in the thermodynamic limit, as the lower bound for the synchronization coupling becomes infinite. The fact that heterogeneity does not always imply better synchronizability, in spite of a significant decrease in network diameter and average distance between pairs of nodes, was, for a long time, known as the *paradox of heterogeneity*.

5.5 Eigenvalue Bounds

A common occurrence in the study of complex networks is the need to analyze a specific system and estimate its properties based on graph-theoretical observables. Quantities such as the minimum and maximum degrees, or the diameter of the network, allow one to compute bounds on the Laplacian eigenvalues, which, in turn, give insight into the synchronizability of a specific network. In the following, we illustrate some of the main graph-theoretical eigenvalue bounds.

5.5.1 Minimum Degree Upper Bound for the Spectral Gap

The smallest nonzero eigenvalue λ_2 is limited by the smallest degree in the network. To prove this, one can study the spectral properties of a carefully constructed matrix and infer the result (Fiedler 1973).

Let $\mathbf{A}$ be a real symmetric matrix. Then, it is

$$\mathbf{V}^{-1}\mathbf{A}\mathbf{V} = \mathbf{D}, \tag{5.68}$$

where the columns of $\mathbf{V}$ are the eigenvectors of $\mathbf{A}$, and $\mathbf{D}$ is the diagonal matrix with the corresponding eigenvalues. Since $\mathbf{A}$ is real and symmetric, it is always possible to choose its eigenvectors to be orthonormal. Thus, $\mathbf{V}$ is orthogonal, and it is $\mathbf{V}^{-1} = \mathbf{V}^{\mathrm{T}}$.

Now, order the eigenvectors so that the eigenvalues are nondecreasing: $\lambda_1 \leqslant \lambda_2 \leqslant \cdots \lambda_N$. Because $\mathbf{D}$ is diagonal, its eigenvectors are the canonical basis $\mathbf{e}_1, \mathbf{e}_2, \ldots, \mathbf{e}_N$, whose elements obey the rule $(\mathbf{e}_i)_j = \delta_{i,j}$. Then, consider a unit vector $\mathbf{x}$, orthogonal to $\mathbf{e}_1$. It is

$$\mathbf{x}^{\mathrm{T}}\mathbf{D}\mathbf{x} = \sum_{i=1}^{N} \lambda_i x_i^2 . \tag{5.69}$$

From $\mathbf{x} \perp \mathbf{e}_1$, it follows that $x_1 = 0$, since the only nonzero element of $\mathbf{e}_1$ is $(\mathbf{e}_1)_1$. Then, the equation above becomes

$$\mathbf{x}^{\mathrm{T}}\mathbf{D}\mathbf{x} = \sum_{i=2}^{N} \lambda_i x_i^2 . \tag{5.70}$$

Since the eigenvalues are nondecreasing, we can write

$$\mathbf{x}^{\mathrm{T}}\mathbf{D}\mathbf{x} \geqslant \lambda_2 \sum_{i=2}^{N} x_i^2 = \lambda_2 , \tag{5.71}$$

where in the equality we used the fact that $\mathbf{x}$ is a unit vector. At the same time,

$$\mathbf{e}_2{}^{\mathrm{T}}\mathbf{D}\mathbf{e}_2 = \lambda_2 . \tag{5.72}$$

Equations 5.71 and 5.72 together imply that

$$\lambda_2 = \min_{\substack{\mathbf{x}\perp\mathbf{e}_1 \\ \|\mathbf{x}\|=1}} \mathbf{x}^{\mathrm{T}}\mathbf{D}\mathbf{x} . \tag{5.73}$$

Substitute Equation 5.68 into the equation above to obtain

$$\lambda_2 = \min_{\substack{\mathbf{x}\perp\mathbf{e}_1 \\ \|\mathbf{x}\|=1}} \mathbf{x}^{\mathrm{T}}\mathbf{V}^{\mathrm{T}}\mathbf{A}\mathbf{V}\mathbf{x} . \tag{5.74}$$

Now, let

$$\mathbf{y} = \mathbf{V}\mathbf{x} , \tag{5.75}$$

or, equivalently, $\mathbf{x} = \mathbf{V}^{\mathrm{T}}\mathbf{y}$. Since $\mathbf{V}$ is an orthogonal matrix, it conserves the metric: $\|\mathbf{V}\mathbf{x}\| = \|\mathbf{x}^{\mathrm{T}}\mathbf{V}^{\mathrm{T}}\| = \|\mathbf{x}\|$.

Thus,

$$\|\mathbf{x}\| = 1 \Rightarrow \|\mathbf{y}\| = 1 . \tag{5.76}$$

Also, if $\mathbf{x} \perp \mathbf{e}_1$, it is $0 = \mathbf{x}^T\mathbf{e}_1 = \mathbf{y}^T\mathbf{V}\mathbf{e}_1 = \mathbf{y}^T\mathbf{v}_1$, where $\mathbf{v}_1$ is the eigenvector of $\mathbf{A}$ corresponding to the first eigenvalue λ_1. Thus,

$$\mathbf{y} \perp \mathbf{v}_1 . \tag{5.77}$$

Substituting Equations 5.75, 5.76 and 5.77 into Equation 5.74, we finally obtain

$$\lambda_2 = \min_{\substack{\mathbf{y}\perp\mathbf{v}_1 \\ \|\mathbf{y}\|=1}} \mathbf{y}^T\mathbf{A}\mathbf{y} . \tag{5.78}$$

Note that the same demonstration can be repeated using a vector $\mathbf{x}$ orthogonal to $\mathbf{e}_N$, rather than $\mathbf{e}_1$. Then, the sums range from 1 to $N - 1$, the inequality in Equation 5.71 has a "less-than-or-equal-to" sign, and the relevant eigenvalue is λ_{N-1}. Continuing the procedure results in

$$\lambda_{N-1} = \max_{\substack{\mathbf{y}\perp\mathbf{v}_N \\ \|\mathbf{y}\|=1}} \mathbf{y}^T\mathbf{A}\mathbf{y} , \tag{5.79}$$

a result that will be useful later.

Equations 5.78 and 5.79 are expressions of the Courant theorem, and, as demonstrated, hold for any real symmetric matrix, including any graph Laplacian $\mathbf{L}$, for which it is $\mathbf{v}_1 = \left(1/\sqrt{N}, 1/\sqrt{N}, \ldots, 1/\sqrt{N}\right)^T$ and $\lambda_1 = 0$. Then, given a Laplacian $\mathbf{L}$, consider the matrix

$$\mathbf{M} = \mathbf{L} - \lambda_2\left(\mathbb{1} - \frac{1}{N}\mathbb{J}\right) , \tag{5.80}$$

where $\mathbb{J}$ is the $N \times N$ matrix whose elements are all 1. To reach our final result, we need to show that $\mathbf{M}$ is positive semidefinite. To see this, first verify that $\mathbf{v}_1$ is also an eigenvector of $\mathbb{J}$ with eigenvalue N, and an eigenvector of $\mathbf{M}$ with eigenvalue 0. In fact, since all the elements of $\mathbb{J}$ are 1, applying it to any vector $\mathbf{w}$ results in a vector whose components are all identical and equal to the sum of the components of $\mathbf{w}$. Then, it is

$$\mathbb{J}\mathbf{v}_1 = \left(\frac{N}{\sqrt{N}}, \frac{N}{\sqrt{N}}, \ldots, \frac{N}{\sqrt{N}}\right)^T = N\mathbf{v}_1 , \tag{5.81}$$

hence

$$\begin{aligned}
\mathbf{M}\mathbf{v}_1 &= \mathbf{L}\mathbf{v}_1 - \lambda_2\left(\mathbb{1}\mathbf{v}_1 - \frac{1}{N}\mathbb{J}\mathbf{v}_1\right) \\
&= \mathbf{0} - \lambda_2\left(\mathbf{v}_1 - \frac{1}{N}N\mathbf{v}_1\right) \\
&= \mathbf{0} .
\end{aligned} \tag{5.82}$$

Now we can prove that, given a vector $\mathbf{z}$, it is always $\mathbf{z}^T\mathbf{Mz} \geqslant 0$. Any vector $\mathbf{z}$ can be decomposed in a linear combination of $\mathbf{v}_1$ and a unit vector $\mathbf{y}$ belonging to the subspace of $\mathbb{R}^N$ orthogonal to $\mathbf{v}_1$:

$$\mathbf{z} = \alpha\mathbf{v}_1 + \beta\mathbf{y}, \tag{5.83}$$

with $\|\mathbf{y}\| = 1$ and $\mathbf{y} \perp \mathbf{v}_1$. Then, it is

$$\mathbf{z}^T\mathbf{Mz} = \alpha^2\mathbf{v}_1^T\mathbf{Mv}_1 + \beta^2\mathbf{y}^T\mathbf{My}$$
$$= \beta^2\left[\mathbf{y}^T\mathbf{Ly} - \lambda_2\left(\mathbf{y}^T\mathbb{1}\mathbf{y} - \frac{1}{N}\mathbf{y}^T\mathbb{J}\mathbf{y}\right)\right]. \tag{5.84}$$

But $\mathbb{J}\mathbf{y} = (S, S, \ldots, S)^T$, where $S = \sum_{i=1}^{N} y_i$. However, $\sum_{i=1}^{N} y_i = \sqrt{N}\mathbf{v}_1^T\mathbf{y} = 0$, since $\mathbf{y} \perp \mathbf{v}_1$. Thus, $\mathbb{J}\mathbf{y} = \mathbf{0}$, hence

$$\mathbf{z}^T\mathbf{Mz} = \beta^2\left(\mathbf{y}^T\mathbf{Lx} - \lambda_2\mathbf{y}^T\mathbf{y}\right)$$
$$= \beta^2\left(\mathbf{y}^T\mathbf{Ly} - \lambda_2\right). \tag{5.85}$$

From Equation 5.78 it follows that $\mathbf{y}^T\mathbf{Ly} \geqslant \lambda_2$, hence $\mathbf{z}^T\mathbf{Mz} \geqslant 0$ and $\mathbf{M}$ is positive semi-definite. In turn, this implies that the diagonal elements of $\mathbf{M}$ are all nonnegative. This is true in particular for the smallest such element:

$$\min_i M_{i,i} = \min_i L_{i,i} - \lambda_2\left(1 - \frac{1}{N}\right) \geqslant 0. \tag{5.86}$$

Since the diagonal elements of a graph Laplacian are the degrees of the corresponding nodes, this inequality can be rewritten as

$$\min_i k_i - \lambda_2\frac{N-1}{N} \geqslant 0, \tag{5.87}$$

from which it follows that

$$\lambda_2 \leqslant \frac{N}{N-1}\min_i k_i. \tag{5.88}$$

This result constitutes a first upper bound on the smallest nonzero eigenvalue in terms of the degrees of the network. Also, it confirms the observation that preferential attachment networks, whose minimum degree is always 1, are hard to synchronize, as their λ_2 can never be too large.

5.5.2 Connectivity Lower Bound on the Spectral Gap

The next result we are going to prove is a lower bound on λ_2, expressed in terms of the edge connectivity of a network. Following Fiedler (1972, 1973), the proof consists of several steps, and it makes use of a few new concepts. We proceed as follows: first, we introduce the measure of irreducibility μ of a matrix, and discuss its conceptual connection with edge connectivity; then, we define the auxiliary matrices $\tilde{\mathbf{P}}$, $\tilde{\mathbf{C}}$ and $\mathbf{U}$, as well as a matrix function $\mathbf{H}(\mu)$; using these, we study an auxiliary matrix family $\mathbf{M}$, which allows us to prove an important property of symmetric stochastic matrices; finally, the result on λ_2 follows from the application of this property to a specific matrix built from the graph Laplacian $\mathbf{L}$.

Irreducibility, Edge Connectivity and Stochastic Matrices

A square matrix $\mathbf{A}$ of dimension N is said to be reducible by definition if it is permutation-similar to a block-triangular matrix. This is equivalent to saying that one can identify a set of integers $\mathcal{N}$, all between 1 and N and with $|\mathcal{N}| < N$, such that $A_{i,j} = 0$ for every $i \in \mathcal{N}$ and $j \notin \mathcal{N}$. To see this, permute $\mathbf{A}$ so that the first $|\mathcal{N}|$ rows and columns are those corresponding to the elements of $\mathcal{N}$. This results in a rectangular block of elements, all above the diagonal, whose value is 0. Conversely, if no such set $\mathcal{N}$ exists, then the matrix is called *irreducible*.

This definition lends itself well to a graph-theoretical interpretation: if one considers $\mathbf{A}$ to be the adjacency matrix of a directed weighted graph, then $\mathbf{A}$ is irreducible if and only if the graph is strongly connected (or simply connected if $\mathbf{A}$ is symmetric and the graph is undirected). This follows easily from the equivalence between a permutation similarity transformation of $\mathbf{A}$ and a relabeling of the nodes of the graph it represents.

Thus, the existence of a set $\mathcal{N}$ as defined above implies the possibility of partitioning the graph into two components G_1 and G_2 such that a path exists at most from all the nodes in G_1 to some node in G_2, but no path exists from any node in G_2 to any node in G_1. Note that the graph is not necessarily simple, as no assumption is made on the diagonal elements of $\mathbf{A}$.

If a matrix is nonnegative, that is, if all its elements are nonnegative, the reducibility condition can be written as

$$\sum_{\substack{i \in \mathcal{N} \\ j \notin \mathcal{N}}} A_{i,j} = 0 . \tag{5.89}$$

Then, given an irreducible nonnegative matrix $\mathbf{A}$, it is natural to introduce a quantity μ that estimates how far $\mathbf{A}$ is from being reducible.

This can be achieved by finding the set $\mathcal{N}$ that minimizes the sum in Equation 5.89. In other words, μ estimates the minimum total of the off-upper-block-triangle elements of $\mathbf{A}$ over all possible permutations. Obviously, $\mathbf{A}$ is reducible if and only if $\mu(\mathbf{A}) = 0$. Equivalently, in terms of network properties, μ measures the minimum total weight of the edges that need to be cut for the graph to separate into two (strongly) connected components. If the network is unweighted, this reduces to the number of edges that need to be cut to achieve this separation, which is the definition of the edge connectivity e of the network.

Throughout this subsection, we are going to make use of results concerning stochastic matrices. By definition, these are nonnegative matrices for which the sum of the elements in each row is 1. It is clear that for such matrices the measure of irreducibility can be at most 1.

Let us now study the set $\mathcal{S}_{\mu^*}$ of stochastic matrices whose measure of irreducibility is greater than or equal to μ^*, and whose properties will be useful later. It is easy to see that this set is closed and bounded, as it contains the stochastic matrices with $\mu = \mu^*$ as well as those with $\mu = 1$. As the existence of such matrices is trivial, the set is also nonempty. Finally, it is also convex.

To prove this last point, consider two matrices $\mathbf{A}_1$ and $\mathbf{A}_2$, and their combination $\mathbf{A}_3 = a\mathbf{A}_1 + b\mathbf{A}_2$, with $a, b \geqslant 0$ and $a + b = 1$. The graph corresponding to $\mathbf{A}_3$ has all the edges of those corresponding to $\mathbf{A}_1$ and $\mathbf{A}_2$. Then, in order to make it *not* strongly connected, one needs to cut at least all the edges that are needed to disconnect the graphs of $\mathbf{A}_1$ and $\mathbf{A}_2$, whose total weight is at least $\mu(\mathbf{A}_1)$ and $\mu(\mathbf{A}_2)$, respectively. Thus

$$\mu(\mathbf{A}_3) \geqslant a\mu(\mathbf{A}_1) + b\mu(\mathbf{A}_2) . \tag{5.90}$$

At the same time, if $\mathbf{A}_1$ and $\mathbf{A}_2$ are in $\mathcal{S}_{\mu^*}$, so is $\mathbf{A}_3$, since $\mu(\mathbf{A}_1)$ and $\mu(\mathbf{A}_2)$ are at least equal to μ^*. So, it is

$$\mu(\mathbf{A}_3) \geqslant a\mu^* + b\mu^* = (a + b)\mu^* = \mu^* , \tag{5.91}$$

which proves that $\mathcal{S}_{\mu^*}$ is also convex.

Auxiliary Quantities and Their Properties

We are now going to introduce several auxiliary quantities that will help us reach our final result. First consider the matrix $\tilde{\mathbf{P}}$ whose elements are

$$\tilde{P}_{i,j} = \frac{1}{2} \left[\delta_{i,j+1} + \delta_{i,j-1} + \delta_{i,j} \left(\delta_{i,1} + \delta_{i,N} \right) \right] . \tag{5.92}$$

In other words, $\tilde{\mathbf{P}}$ is $1/2$ times the adjacency matrix of a path graph on N nodes with loops on the ends (Figure 5.7):

$$\tilde{\mathbf{P}} = \frac{1}{2}\begin{pmatrix} 1 & 1 & 0 & \cdots & 0 & 0 & 0 \\ 1 & 0 & 1 & \ddots & 0 & 0 & 0 \\ 0 & 1 & 0 & \ddots & 0 & 0 & 0 \\ \vdots & \ddots & \ddots & \ddots & \ddots & \ddots & \vdots \\ 0 & 0 & 0 & \ddots & 0 & 1 & 0 \\ 0 & 0 & 0 & \ddots & 1 & 0 & 1 \\ 0 & 0 & 0 & \cdots & 0 & 1 & 1 \end{pmatrix} . \tag{5.93}$$

Figure 5.7 Path graph on N nodes with loops on the ends, whose adjacency matrix is used in the definition of $\tilde{\mathbf{P}}$, Equation 5.93.

We are interested in finding the spectrum of $\tilde{\mathbf{P}}$, and in particular its second-largest eigenvalue π_{N-1}. To do so, note that the end loops effectively turn the path on N nodes in Figure 5.7 into a loop on $2N$ nodes, C_{2N}, implying that the eigenvalues of $\tilde{\mathbf{P}}$ are determined by those of C_{2N}.

These latter are easily found by exploiting the fact that the adjacency matrix of a cycle is circulant, and applying the same technique we used in Section 5.3, which yields

$$\chi_t = 2\cos\left((2N - t)\frac{\pi}{N}\right), \quad t \in \mathbb{N}, 1 \leqslant t \leqslant 2N. \tag{5.94}$$

Since the graph is an improper cycle, in which each node is touched twice, the spectrum of $\tilde{\mathbf{P}}$ retains one of each pair of degenerate eigenvalues of C_{2N}. Thus, it is

$$\pi_t = \cos\left((N - t)\frac{\pi}{N}\right), \quad t \in \mathbb{N}, 1 \leqslant t \leqslant N, \tag{5.95}$$

where the factor of 2 is canceled by the $1/2$ in the definition of $\tilde{\mathbf{P}}$.

Define now the matrix function

$$\mathbf{H}(\mu) = (1 - 2\mu)\mathbb{1} + 2\mu\tilde{\mathbf{P}}. \tag{5.96}$$

Once more, we are interested in the second largest eigenvalue $\eta_{N-1}(\mu)$ of $\mathbf{H}$. From the definition, it is

$$\eta_{N-1}(\mu) = 1 - 2\mu + 2\mu\pi_{N-1}$$
$$= 1 - 2\mu\left(1 - \cos\frac{\pi}{N}\right). \tag{5.97}$$

Finally, let $\mathbf{U}$ be the upper-triangular matrix-of-ones, whose elements $U_{i,j}$ are 1 if $i \leqslant j$ and 0 otherwise:

$$\mathbf{U} = \begin{pmatrix} 1 & 1 & \cdots & 1 \\ 0 & 1 & \cdots & 1 \\ \vdots & \ddots & \ddots & \vdots \\ 0 & 0 & \cdots & 1 \end{pmatrix}. \tag{5.98}$$

One More Nonnegative Matrix

We can now show that, for every symmetric stochastic matrix $\mathbf{A}$ such that $\mu(\mathbf{A}) \geqslant \mu^*$, the elements of the matrix

$$\mathbf{M} = \mathbf{U}^{\mathrm{T}} \left(\mathbf{H} \left(\mu^* \right) - \mathbf{A} \right) \mathbf{U} \tag{5.99}$$

are all nonnegative and, in addition, it is $M_{i,N} = M_{N,i} = 0$ for all i.

To see this, first express the elements of $\mathbf{M}$ in terms of the row-by-column product of the matrices in Equation 5.99. The (i, j) element of $\mathbf{M}$ is the inner product of the ith row of $\mathbf{U}^{\mathrm{T}}$ with the jth column of $(\mathbf{H}(\mu^*) - \mathbf{A})\mathbf{U}$. In turn, this is the product of $\mathbf{H}(\mu^*) - \mathbf{A}$ and the jth column of $\mathbf{U}$, so that

$$\begin{aligned} M_{i,j} &= \mathbf{U_i}^{\mathrm{T}} \left(\mathbf{H} \left(\mu^* \right) - \mathbf{A} \right) \mathbf{U_j} \\ &= \mathbf{U_i}^{\mathrm{T}} \mathbf{H} \left(\mu^* \right) \mathbf{U_j} - \mathbf{U_i}^{\mathrm{T}} \mathbf{A} \mathbf{U_j} , \end{aligned} \tag{5.100}$$

where $\mathbf{U_i}$ is the ith column of $\mathbf{U}$ and we used the fact that the ith row of $\mathbf{U}^{\mathrm{T}}$ is the ith column of $\mathbf{U}$.

Then, to verify that all the $M_{i,j} \geqslant 0$, start by computing the first term in Equation 5.100. Using the definition of $\mathbf{H}$, Equation 5.96, it is

$$\mathbf{U_i}^{\mathrm{T}} \mathbf{H} \left(\mu^* \right) \mathbf{U_j} = \left(1 - 2\mu^* \right) \mathbf{U_i}^{\mathrm{T}} \mathbb{1} \mathbf{U_j} + 2\mu^* \mathbf{U_i}^{\mathrm{T}} \tilde{\mathbf{P}} \mathbf{U_j} . \tag{5.101}$$

Since all the matrices are symmetric, we will consider only the upper triangle $j \geqslant i$. The vector $\mathbf{U_i}$ is a vector whose first i elements are 1 and the rest are 0. Then, the term $\mathbf{U_i}^{\mathrm{T}} \tilde{\mathbf{P}} \mathbf{U_j}$ is the sum of the elements in the first j columns and the first i rows of $\tilde{\mathbf{P}}$. Thus, from Equation 5.93, it is

$$\mathbf{U_i}^{\mathrm{T}} \tilde{\mathbf{P}} \mathbf{U_j} = i - \frac{1}{2} , \tag{5.102}$$

hence

$$\mathbf{U_i}^{\mathrm{T}} \mathbf{H} \left(\mu^* \right) \mathbf{U_j} = \begin{cases} i - 2\mu^* i + 2\mu^* i - \mu^* = i - \mu^* & \text{for } j = i \\ i - 2\mu^* i + 2\mu^* i \quad\quad = i & \text{for } j > i , \end{cases} \tag{5.103}$$

from which it follows that

$$M_{i,j} = \begin{cases} i - \mu^* - \mathbf{U_i}^{\mathrm{T}} \mathbf{A} \mathbf{U_j} & \text{for } j = i \\ i - \mathbf{U_i}^{\mathrm{T}} \mathbf{A} \mathbf{U_j} & \text{for } j > i . \end{cases} \tag{5.104}$$

Summing and subtracting the vector-of-ones $\mathbf{1}$, it is

$$\begin{aligned} \mathbf{U_i}^{\mathrm{T}} \mathbf{A} \mathbf{U_j} &= \mathbf{U_i}^{\mathrm{T}} \mathbf{A} \left(1 - 1 + \mathbf{U_j} \right) \\ &= \mathbf{U_i}^{\mathrm{T}} \mathbf{A} \left[1 - \left(1 - \mathbf{U_j} \right) \right] \\ &= \mathbf{U_i}^{\mathrm{T}} \mathbf{A} \mathbf{1} - \mathbf{U_i}^{\mathrm{T}} \mathbf{A} \left(1 - \mathbf{U_j} \right) \\ &= \mathbf{U_i}^{\mathrm{T}} \mathbf{1} - \mathbf{U_i}^{\mathrm{T}} \mathbf{A} \left(1 - \mathbf{U_j} \right) \\ &= i - \mathbf{U_i}^{\mathrm{T}} \mathbf{A} \left(1 - \mathbf{U_j} \right) , \end{aligned} \tag{5.105}$$

where we used the fact that, for any stochastic matrix, $\mathbf{1}$ is an eigenvector with eigenvalue 1. From here, it is

$$\mathbf{U_i}^{\mathrm{T}} \mathbf{A} \left(1 - \mathbf{U_j} \right) = i - \mathbf{U_i}^{\mathrm{T}} \mathbf{A} \mathbf{U_j} . \tag{5.106}$$

But $U_i{}^T A (1 - U_j)$ is a sum of elements of A, and, as such, it is nonnegative. Thus,

$$i - U_i{}^T A U_j \geqslant 0 \Rightarrow$$
$$U_i{}^T A U_j \leqslant i ,$$

(5.107)

which, together with Equation 5.104, implies that the off-diagonal terms of M are all nonnegative.

Consider now Equation 5.105 when $j = i$. The first i elements of the vector $1 - U_i$ are all 0, while the remaining ones are all 1. Thus, the term $U_i{}^T A (1 - U_i)$ is the sum of the elements in the first i rows and the last $N - i$ columns of A, that is, the sum of some off-block-diagonal elements of this matrix. By the definition of measure of irreducibility, this is at least as big as $\mu (A)$, which is in turn at least μ^* by hypothesis. Then, from Equation 5.105, it is

$$U_i{}^T A U_j \leqslant i - \mu^* ,$$

(5.108)

which, with Equation 5.104, implies that all $M_{i,i} \geqslant 0$,

In addition, $U_N = 1$. Thus, exploiting the fact that $\tilde{P}$ is a symmetric stochastic matrix,

$$\begin{aligned}
M_{i,N} &= U_i{}^T H (\mu^*) U_N - U_i{}^T A U_N \\
&= U_i{}^T H (\mu^*) 1 - U_i{}^T A 1 \\
&= (1 - 2\mu^*) U_i{}^T 11 + 2\mu^* U_i{}^T \tilde{P} 1 - U_i{}^T 1 \\
&= i (1 - 2\mu^*) + 2\mu^* i - 1 \\
&= 0 ,
\end{aligned}$$

(5.109)

which completes the proof.

A Property of Symmetric Stochastic Matrices

Before reaching our final result, we need to prove one more statement: if A is a symmetric stochastic matrix, with eigenvalues $\alpha_1 \leqslant \alpha_2 \leqslant \cdots \leqslant \alpha_N = 1$, then $\alpha_{N-1} \leqslant \eta_{N-1} (\mu (A))$. To prove this, the first step is showing that

$$\max_{A \in \mathcal{S}_{\mu^*}} \alpha_{N-1} = \eta_{N-1} (\mu^*) .$$

(5.110)

The properties of $\mathcal{S}_{\mu^*}$ proved above guarantee the existence of the maximum. Then, applying Equation 5.79, it is

$$\begin{aligned}
\max_{A \in \mathcal{S}_{\mu^*}} \alpha_{N-1} &= \max_{A \in \mathcal{S}_{\mu^*}} \max_{\substack{x \perp 1 \\ \|x\|=1}} x^T A x \\
&= \max_{\substack{x \perp 1 \\ \|x\|=1}} \max_{A \in \mathcal{S}_{\mu^*}} x^T A x .
\end{aligned}$$

(5.111)

In considering the equation above, we can restrict ourselves to vectors $\mathbf{x}$ whose elements are nonincreasing, since permutations do not affect the measure of irreducibility. Then, for any such vector $\mathbf{x}$, create the vector $\mathbf{y} = \mathbf{U}^{-1}\mathbf{x}$. For any $\mathbf{A} \in \mathcal{S}_{\mu^*}$, it is

$$
\begin{aligned}
\mathbf{y}^{\mathrm{T}}\mathbf{M}\mathbf{y} &= \mathbf{x}^{\mathrm{T}} \left(\mathbf{U}^{-1}\right)^{\mathrm{T}} \mathbf{U}^{\mathrm{T}} \left(\mathbf{H}\left(\mu^*\right) - \mathbf{A}\right) \mathbf{U}\mathbf{U}^{-1}\mathbf{x} \\
&= \mathbf{x}^{\mathrm{T}} \left(\mathbf{U}\mathbf{U}^{-1}\right)^{\mathrm{T}} \left(\mathbf{H}\left(\mu^*\right) - \mathbf{A}\right) \mathbf{U}\mathbf{U}^{-1}\mathbf{x} \qquad\qquad (5.112) \\
&= \mathbf{x}^{\mathrm{T}}\mathbf{H}\left(\mu^*\right)\mathbf{x} - \mathbf{x}^{\mathrm{T}}\mathbf{A}\mathbf{x} .
\end{aligned}
$$

But the elements of $\mathbf{y}$ are all nonnegative except, at most, the last. In fact, the matrix $\mathbf{U}^{-1}$ has all the diagonal elements equal to 1, all the superdiagonal elements equal to -1, and all the remaining equal to 0:

$$
\mathbf{U}^{-1} = \begin{pmatrix}
1 & -1 & 0 & \cdots & 0 & 0 & 0 \\
0 & 1 & -1 & \ddots & 0 & 0 & 0 \\
0 & 0 & 1 & \ddots & 0 & 0 & 0 \\
\vdots & \ddots & \ddots & \ddots & \ddots & \ddots & \vdots \\
0 & 0 & 0 & \ddots & 1 & -1 & 0 \\
0 & 0 & 0 & \ddots & 0 & 1 & -1 \\
0 & 0 & 0 & \cdots & 0 & 0 & 1
\end{pmatrix} . \qquad\qquad (5.113)
$$

Thus, for all $i < N$ it is $y_i = x_i - x_{i+1} \geqslant 0$, since the elements of $\mathbf{x}$ are non-increasing. But then it is also $\mathbf{y}^{\mathrm{T}}\mathbf{M}\mathbf{y} \geqslant 0$, since all the elements of $\mathbf{M}$ are non-negative and its last row and column consist entirely of zeros, as proved above.

It follows that

$$
\mathbf{x}^{\mathrm{T}}\mathbf{H}\left(\mu^*\right)\mathbf{x} - \mathbf{x}^{\mathrm{T}}\mathbf{A}\mathbf{x} \geqslant 0 , \qquad\qquad (5.114)
$$

hence

$$
\mathbf{x}^{\mathrm{T}}\mathbf{A}\mathbf{x} \leqslant \mathbf{x}^{\mathrm{T}}\mathbf{H}\left(\mu^*\right)\mathbf{x} . \qquad\qquad (5.115)
$$

Note that $\mathbf{H}$ is in general not a stochastic matrix. In fact, as it is evident from its definition, Equation 5.96, some of its elements become negative when $\mu > 1/2$. Nonetheless, the elements in each row still sum to 1, even in this case. Thus, $\mathbf{1}$ is an eigenvector of $\mathbf{H}$ with eigenvalue 1. Since Equation 5.97 implies that its second-largest eigenvalue is at most 1, then 1 is its largest eigenvalue. Then, taking the maximum over all the unit vectors orthogonal to $\mathbf{1}$ yields

$$
\alpha_{N-1} \leqslant \eta_{N-1}\left(\mu^*\right) . \qquad\qquad (5.116)
$$

This holds for all $\mathbf{A} \in \mathcal{S}_{\mu^*}$, and in particular for the $\mathbf{A}$ that maximizes the left-hand side. Thus

$$\max_{\mathbf{A} \in \mathcal{S}_{\mu^*}} \alpha_{N-1} = \eta_{N-1}\left(\mu^*\right) , \tag{5.117}$$

which is an equality because $\mathcal{S}_{\mu^*}$ is closed and bounded, and implies that

$$\alpha_{N-1} \leqslant \eta_{N-1}\left(\mu\left(\mathbf{A}\right)\right) . \tag{5.118}$$

Lower Bound on λ_2

We are now ready for our final result. Given a Laplacian $\mathbf{L}$, consider the matrix

$$\mathbf{S} = \mathbb{1} - \frac{\mathbf{L}}{\max_i k_i} . \tag{5.119}$$

This is a symmetric stochastic matrix. Thus, from Equation 5.118 it follows that its second largest eigenvalue is

$$\sigma_{N-1} \leqslant \eta_{N-1}\left(\mu\left(\mathbf{S}\right)\right) , \tag{5.120}$$

hence

$$\sigma_{N-1} \leqslant 1 - 2\mu\left(\mathbf{S}\right)\left(1 - \cos\frac{\pi}{N}\right) . \tag{5.121}$$

But $\mu\left(\mathbf{S}\right)$ is just the edge connectivity of the network divided by its largest degree, and

$$\sigma_{N-1} = 1 - \frac{\lambda_2}{\max_i k_i} . \tag{5.122}$$

Then, it is

$$1 - \frac{\lambda_2}{\max_i k_i} \leqslant 1 - \frac{2e}{\max_i k_i}\left(1 - \cos\frac{\pi}{N}\right) , \tag{5.123}$$

hence

$$\lambda_2 \geqslant 2e\left(1 - \cos\frac{\pi}{N}\right) , \tag{5.124}$$

which is the result we wanted to prove.

5.5.3 Diameter Lower Bound for the Spectral Gap

Among the various structural features of a network, the diameter D offers one more estimate for a bound on λ_2. By definition

$$D = \max_{i,j} d\left(i, j\right) , \tag{5.125}$$

where $d\,(i,j)$ is the distance between node i and node j, that is, the minimum length of any path connecting i and j. The result on λ_2 is due to a theorem by Mohar and McKay (1991), which makes use of the properties of shortest paths.

To derive the result, one must first prove an upper bound on the number of shortest paths in a graph G to which a given edge can belong. In particular, if one chooses one possible shortest path $P_{i,j}$ for all pairs of nodes i and j, then any given edge can belong to at most $N^2/4$ of the chosen paths. To see this, alter choosing the paths, create a family of graphs G_ℓ, where ℓ is an edge of G. Every G_ℓ has the same nodes as G, but every pair of nodes i and j has a link if and only if the edge ℓ belongs to the shortest path $P_{i,j}$ chosen.

We want to show that the maximum number of edges in any G_ℓ is $N^2/4$. We prove this by showing that G_ℓ is locally tree-like, which we prove by contradiction. Assume the three nodes i, j, and k form a triangle in G_ℓ. This means that the edge ℓ belongs to the shortest paths $P_{i,j}$, $P_{i,k}$, and $P_{j,k}$. The edge ℓ is traversed in the same direction along two of these paths. Without loss of generality, let these be $P_{i,k}$ and $P_{j,k}$, and let ℓ be formed by the nodes x and y, traversed in this order. Then it is

$$d\,(i,y) > d\,(i,x)$$
$$d\,(j,y) > d\,(j,x)\,. \qquad (5.126)$$

Also, by the triangle inequality,

$$d\,(i,j) \leqslant d\,(i,x) + d\,(x,j)\,. \qquad (5.127)$$

Using the inequalities in Equation 5.126 into the inequality above, one obtains

$$d\,(i,j) < d\,(i,x) + d\,(j,y)$$
$$< d\,(i,x) + 1 + d\,(j,y) \qquad (5.128)$$
$$= d\,(i,x) + d\,(x,y) + d\,(j,y)$$

and

$$d\,(i,j) < d\,(i,y) + d\,(x,j)$$
$$< d\,(i,y) + 1 + d\,(x,j) \qquad (5.129)$$
$$= d\,(i,y) + d\,(y,x) + d\,(x,j)\,.$$

Thus, the distance between i and j is always less than the length of any path from i to j that includes the edge ℓ, which, therefore, cannot belong to the G_ℓ considered, contradicting the hypothesis. This proves that any G_ℓ is triangle-free.

In addition, any triangle-free graph can have at most $N^2/4$ edges. We prove this by induction. The statement is obviously true for $N = 1$ and $N = 2$. Then, we assume it is valid for $N - 2$, and prove this implies validity for N. Start by considering a graph G with two linked nodes i and j and the graph $\tilde{G}$ containing all the nodes of G except i and j. The number of edges E of G_ℓ is

$$E = \tilde{E} + k_i + k_j - 1 , \tag{5.130}$$

where $\tilde{E}$ is the number of edges of $\tilde{G}$. But then, by induction hypothesis, it is

$$\tilde{E} \leqslant \frac{(N - 2)^2}{4} , \tag{5.131}$$

since $\tilde{E}$ has $N - 2$ nodes. Also, $k_i + k_j$ cannot be greater than N, since there are no triangles in G_ℓ, Then, one has

$$
\begin{aligned}
\tilde{E} + k_i + k_j - 1 &\leqslant \frac{(N - 2)^2}{4} + N - 1 \\
&= \frac{N^2 + 4 - 4N}{4} + N - 1 \\
&- \frac{N^2}{4} ,
\end{aligned}
\tag{5.132}
$$

hence $E \leqslant N^2/4$, which is what we wanted to prove.

We can now prove the lower bound on λ_2. Let $\mathbf{v}$ be the eigenvector corresponding to λ_2. The sum of the squared differences of its elements over all pairs of nodes is

$$
\begin{aligned}
\sum_{i=1}^{N} \sum_{j=1}^{N} \left(v_i - v_j \right)^2 &= \sum_{i=1}^{N} \sum_{j=1}^{N} \left(v_i^2 + v_j^2 - 2 v_i v_j \right) \\
&= N \sum_{i=1}^{N} v_i^2 + N \sum_{j=1}^{N} v_j^2 - 2 \sum_{i=1}^{N} \sum_{j=1}^{N} v_i v_j \\
&= 2N - 2 \sum_{i=1}^{N} \left(v_i \sum_{j=1}^{N} v_j \right) \\
&= 2N ,
\end{aligned}
\tag{5.133}
$$

where we used the fact that $\mathbf{v}$ is a unit vector and that the sum of its elements vanishes because it is orthogonal to $\mathbf{1}$. At the same time, each term $\left(v_i - v_j \right)^2$ in the sum can be expanded by summing and subtracting all the elements corresponding to the nodes along the shortest path $P_{i,j} = \left(i, p_1, p_2, \ldots, p_{d(i,j)-1}, j \right)$:

$$
\begin{aligned}
\left(v_i - v_j \right)^2 &= \left(v_i - v_{p_1} + v_{p_1} - v_{p_2} + v_{p_2} - \cdots - v_{p_{d(i,j)-1}} + v_{p_{d(i,j)-1}} - v_j \right)^2 \\
&= \left[\left(v_i - v_{p_1} \right) + \left(v_{p_1} - v_{p_2} \right) + \cdots + \left(v_{p_{d(i,j)-1}} - v_j \right) \right]^2 .
\end{aligned}
\tag{5.134}
$$

Introducing the notation $\Delta^2 (\ell) = \left(v_x - v_y \right)^2$, where ℓ is an edge between node x and node y, it is

$$\left(v_i - v_j\right)^2 \leqslant d\,(i,\,j)\left[\left(v_i - v_{p_1}\right)^2 + \left(v_{p_1} - v_{p_2}\right)^2 + \cdots + \left(v_{p_{d(i,j)-1}} - v_j\right)^2\right]$$

$$= d\,(i,\,j)\sum_{\ell \in P_{i,j}} \Delta^2\,(\ell). \tag{5.135}$$

The sum over all edges of $P_{i,j}$ can be extended to a sum over all edges of the network by defining the function $I_{i,j}\,(\ell)$ to be 1 if $\ell \in P_{i,j}$ and 0 otherwise. Then, one obtains

$$\left(v_i - v_j\right)^2 \leqslant d\,(i,\,j)\sum_{\ell \in G} \Delta^2\,(\ell)\,I_{i,j}\,(\ell)\;. \tag{5.136}$$

Summing both sides of the inequality over all i and j, it is

$$\sum_{i=1}^{N}\sum_{j=1}^{N}\left(v_i - v_j\right)^2 \leqslant \sum_{i=1}^{N}\sum_{j=1}^{N}d\,(i,\,j)\sum_{\ell \in G} \Delta^2\,(\ell)\,I_{i,j}\,(\ell)$$

$$= \sum_{\ell \in G}\Delta^2\,(\ell)\sum_{i=1}^{N}\sum_{j=1}^{N}d\,(i,\,j)\,I_{i,j}\,(\ell)\;. \tag{5.137}$$

The distance $d\,(i,\,j)$ between i and j is at most as big as the diameter D. Also, we know that a fixed edge ℓ can only appear in at most $N^2/4$ shortest paths, which implies that

$$\sum_{i=1}^{N}\sum_{j=1}^{N}I_{i,j}\,(\ell) \leqslant 2\frac{N^2}{4} = \frac{N^2}{2}\;. \tag{5.138}$$

Then, the inequality becomes

$$\sum_{i=1}^{N}\sum_{j=1}^{N}\left(v_i - v_j\right)^2 \leqslant D\sum_{\ell \in G}\Delta^2\,(\ell)\sum_{i=1}^{N}\sum_{j=1}^{N}I_{i,j}\,(\ell)$$

$$\leqslant \frac{DN^2}{2}\sum_{\ell \in G}\Delta^2\,(\ell)\;. \tag{5.139}$$

Substituting Equation 5.133, it is

$$2N \leqslant \frac{DN^2}{2}\sum_{\ell \in G}\Delta^2\,(\ell)\;, \tag{5.140}$$

hence

$$4 \leqslant DN\sum_{\ell \in G}\Delta^2\,(\ell)\;. \tag{5.141}$$

Writing down $\Delta^2(\ell)$ explicitly in terms of the adjacency matrix, one obtains

$$\sum_{\ell \in G} \Delta^2(\ell) = \sum_{\ell \in G} \left(v_i - v_j\right)^2$$

$$= \frac{1}{2} \sum_{i=1}^{N} \sum_{j=1}^{N} A_{i,j} \left(v_i - v_j\right)^2$$

$$= \frac{1}{2} \left(\sum_{i=1}^{N} \sum_{j=1}^{N} A_{i,j} v_i^2 + \sum_{i=1}^{N} \sum_{j=1}^{N} A_{i,j} v_j^2 - 2 \sum_{i=1}^{N} \sum_{j=1}^{N} A_{i,j} v_i v_j \right)$$

$$= \sum_{i=1}^{N} \sum_{j=1}^{N} A_{i,j} v_i^2 - \sum_{i=1}^{N} \sum_{j=1}^{N} A_{i,j} v_i v_j$$

$$= \sum_{i=1}^{N} v_i^2 \sum_{j=1}^{N} A_{i,j} - \sum_{i=1}^{N} \sum_{j=1}^{N} A_{i,j} v_i v_j$$

$$= \sum_{i=1}^{N} v_i^2 k_i - \sum_{i=1}^{N} v_i \sum_{j=1}^{N} A_{i,j} v_j \qquad (5.142)$$

$$= \sum_{i=1}^{N} v_i \left(k_i v_i - \sum_{j=1}^{N} A_{i,j} v_j \right)$$

$$= \sum_{i=1}^{N} v_i \left(\mathbf{L}\mathbf{v}\right)_i$$

$$= \sum_{i=1}^{N} v_i \lambda_2 v_i$$

$$= \lambda_2 \sum_{i=1}^{N} v_i^2$$

$$= \lambda_2 .$$

And, therefore,

$$4 \leqslant DN\lambda_2 , \qquad (5.143)$$

from which, one finally establishes that

$$\lambda_2 \geqslant \frac{4}{DN} . \qquad (5.144)$$

5.5.4 Degree Bounds on λ_N

With the results derived in the above subsection one can prove some simple bounds on the largest eigenvalue λ_N (Fiedler 1973). First, an upper bound follows easily from the fact that, given a real symmetric matrix $\mathbf{A}$ and a matrix norm $\|\cdot\|$, the largest eigenvalue is always at most as large as the norm. Then, define the row-vector norm

$$\|\mathbf{L}\| = \max_i \sum_{j=1}^{N} \left| L_{i,j} \right| . \tag{5.145}$$

By definition of the graph Laplacian, it is $\|\mathbf{L}\| = 2 \max_i k_i$. Thus,

$$\lambda_N \leqslant 2 \max_i k_i . \tag{5.146}$$

To find the lower bound, consider the Laplacian $\tilde{\mathbf{L}}$ of the complement network, that is, the graph containing all and only the edges not found in the network being studied. This matrix is

$$\tilde{\mathbf{L}} = N\mathbb{1} - \mathbb{J} - \mathbf{L} \tag{5.147}$$

and we know from Equation 5.88 that

$$\tilde{\lambda}_2 \leqslant \frac{N}{N-1} \min_i \tilde{k}_i , \tag{5.148}$$

where $\tilde{\lambda}_i$ are the eigenvalues of $\tilde{\mathbf{L}}$ and $\tilde{k}_i$ are the degrees of the complement network. But, from Equation 5.78, it is

$$
\begin{aligned}
\tilde{\lambda}_2 &= \min_{\substack{\|\mathbf{x}\|=1 \\ \mathbf{x}\perp\mathbb{1}}} \mathbf{x}^{\mathrm{T}} \tilde{\mathbf{L}} \mathbf{x} \\
&= \min_{\substack{\|\mathbf{x}\|=1 \\ \mathbf{x}\perp\mathbb{1}}} \mathbf{x}^{\mathrm{T}} (N\mathbb{1} - \mathbb{J} - \mathbf{L}) \mathbf{x} \\
&= N - \max_{\substack{\|\mathbf{x}\|=1 \\ \mathbf{x}\perp\mathbb{1}}} \mathbf{x}^{\mathrm{T}} \mathbf{L} \mathbf{x} \\
&= N - \lambda_N ,
\end{aligned}
\tag{5.149}
$$

where we used the fact that all eigenvectors of $\mathbb{J}$ except $\mathbb{1}$ correspond to the null eigenvalue, and that the set of eigenvectors of $\mathbf{L}$ forms an orthonormal basis of $\mathbb{R}^N$. Then, substituting Equation 5.149 into Equation 5.148, one obtains

$$N - \lambda_N \leqslant \frac{N}{N-1} \min_i \tilde{k}_i . \tag{5.150}$$

But the smallest degree in the complement network is $N - 1 - \max_i k_i$, so

$$N - \lambda_N \leqslant \frac{N}{N-1} \left(N - 1 - \max_i k_i \right) , \tag{5.151}$$

hence

$$\lambda_N \geqslant N - \frac{N}{N-1}\left(N - 1 - \max_i k_i\right)$$

$$= N - N + \frac{N}{N-1}\max_i k_i \qquad\qquad (5.152)$$

$$= \frac{N}{N-1}\max_i k_i \ .$$

5.5.5 *Summary*

Summarizing the results, we have proved a general upper bound for the spectral gap expressed in terms of the minimum degree of the network,

$$\lambda_2 \leqslant \frac{N}{N-1}\min_i k_i \ , \qquad\qquad (5.88 \text{ rep.})$$

and two lower bounds, functions of the edge connectivity and of the diameter of the network, respectively:

$$\lambda_2 \geqslant 2e\left(1 - \cos\frac{\pi}{N}\right) \qquad\qquad (5.124 \text{ rep.})$$

$$\lambda_2 \geqslant \frac{4}{DN} \ . \qquad\qquad (5.144 \text{ rep.})$$

For the largest eigenvalue λ_N, instead, we have proved an upper bound

$$\lambda_N \leqslant 2\max_i k_i \qquad\qquad (5.146 \text{ rep.})$$

and a lower bound

$$\lambda_N \geqslant \frac{N}{N-1}\max_i k_i \ , \qquad\qquad (5.152 \text{ rep.})$$

both functions of the largest degree of the network.

Clearly, Equation 5.88 converges to the minimum degree for very large N, while the limiting behaviors of Equation 5.124 and Equation 5.144 depend on the scaling properties of the edge connectivity and of the diameter, respectively. Thus, in the thermodynamic limit, for preferential attachment networks, Equation 5.124 becomes

$$\lambda_2 \gtrsim \frac{\pi^2}{N^2} \ , \qquad\qquad (5.153)$$

since their minimum degree is always 1, thus making their edge connectivity 1 as well. Their diameter is known to be quite small, and its scaling depends on their power-law exponent γ (Cohen and Havlin 2003):

$$D_{PA} \sim \begin{cases} \log\left(\log\left(N\right)\right) & \forall\, 2 < \gamma < 3 \\ \frac{\log(N)}{\log(\log(N))} & \text{for } \gamma = 3 \\ \log\left(N\right) & \forall\, \gamma > 3 \ . \end{cases} \qquad (5.154)$$

The bounds obtained substituting the equation above into Equation 5.144 are actually much better than the general one given by the edge connectivity, as shown in Figure 5.8. Similarly, studying the largest eigenvalue λ_N one can see that the actual data are much closer to their theoretical upper bound than they are to the lower one, particularly for higher values of γ (see Figure 5.9).

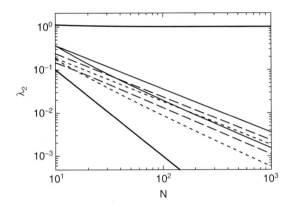

Figure 5.8 Smallest nonzero Laplacian eigenvalue and its bounds for preferential attachment networks. The average spectral gaps are shown for PA networks with $\gamma = 2.5$ (solid lines), $\gamma = 3$ (dashed lines), and $\gamma = 3.5$ (dotted lines), with the corresponding lower bounds computed according to Equations 5.144 and 5.154. The thicker solid line below all the others is the general lower bound of Equation 5.124, while the line at $\lambda_2 \approx 1$ is the upper bound of Equation 5.88. The data are the same as in Figure 5.6.

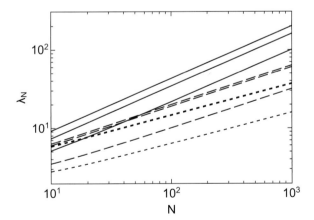

Figure 5.9 Largest Laplacian eigenvalue λ_N and its bounds for preferential attachment networks. The average values of λ_N are shown for PA networks with $\gamma = 2.5$ (solid lines), $\gamma = 3$ (dashed lines), and $\gamma = 3.5$ (dotted lines), together with the corresponding lower and upper bounds (Equation 5.152 and Equation 5.146, respectively). The data are the same as in Figure 5.6. Notice how the actual data are very close to their upper bounds, especially for higher exponents.

5.6 Enhancing and Optimizing Synchronization of Complex Networks

After a general discussion on the properties of the Laplacian spectra for the most common types of networks, the next step is to show briefly how the Master Stability Function approach can be used to enhance and optimize synchronization.

The basic way to accomplish this is a weighting mechanisms whose explicit aim is to reduce the value of the eigenratio λ_N/λ_2 as much as possible.

The first step in this direction was made by Motter et al. (2005), who considered a weighting factor for the coupling term of node i proportional to σ/k_i^β, i.e., inversely proportional to a power $\beta > 0$ of its degree. They demonstrated that an optimal condition $\beta = 1$ for synchronization is found for a large class of networks. This corresponds to a situation where the input strength of the coupling is equal for all nodes. The relevant result is therefore that an improvement of the propensity for synchronization can be obtained with a weighting procedure that retains information on the local features of the network, namely the node degrees.

Weighting procedures based on global information can further enhance the synchronizability of networks. This was demonstrated by Chavez et al. (2005), who considered nodes evolving according to

$$\dot{\mathbf{x}}_i = \mathbf{f}(\mathbf{x}_i) - \frac{\sigma}{\sum_{j\in N_i} \mathcal{L}_{ij}^\alpha} \sum_{j\in N_i} \mathcal{L}_{ij}^\alpha \left(\mathbf{g}(\mathbf{x}_j) - \mathbf{g}(\mathbf{x}_i)\right), \qquad (5.155)$$

where α is a real tunable parameter, N_i is the set of neighbors of node i, and $\mathcal{L}_{ij}$ is the load of the link connecting nodes i and j, which quantifies the traffic on all the shortest paths including that link, thereby reflecting the network structure at a global scale.

More precisely, for each pair of nodes i and j, one counts the number $n(i, j)$ of shortest paths connecting them. For each shortest path, one adds $1/n$ to the load of each link forming it. The load distribution retains information on the network structure at a global level because the value of each load is influenced also by pairs of nodes that are far from either of nodes i and j.

The limit $\alpha = 0$ in Equation 5.155 yields the same condition studied by Motter et al. (2005), since $\sum_{j\in N_i} \mathcal{L}_{ij}^{\alpha=0} = k_i$. The limit $\alpha = +\infty(-\infty)$ induces instead a unidirected N-link tree structure in the wiring, so that only the link with maximum (minimum) load is selected as the incoming link for each node.

By varying α, Chavez et al. (2005) demonstrated that the eigenratio can be made even smaller than that corresponding to $\alpha = 0$, yielding a further improvement in synchronizability. This represents a remarkable result, as it indicates that using global information for a weighting procedure may yield, in general, a better criterion for synchronization than relying solely on the local connectivity information around a node.

After such two first steps, extensions of the MSF approach were created for several cases, including asymmetrically (and even unidirectionally) coupled systems, non-diagonalizable coupling matrices, and almost-identical units. Though being highly relevant, as they show how general the Master Stability Function predictions can be, all these extensions are somehow very technical, and discussing each one of them is certainly beyond the scope of this book. The interested reader is therefore referred to the many papers and reports on the subject that have appeared in the literature since 2005.

5.7 Explosive Synchronization in Complex Networks

Another popular field of research is explosive synchronization, i.e., the induction of a synchronized state by means of a first-order-like transition, in which an order parameter undergoes a generally nonreversible and discontinuous transition at a critical value of the control parameter.

To illustrate the case, let us start by referring to the abrupt transition to synchrony reported by Gómez-Gardeñes et al. (2011), who considered an unweighted and undirected network of N Kuramoto oscillators whose natural frequencies were chosen to be identical to their degree. Their results show that synchronization occurs in a different way for forward and backward transitions. In particular, when the network is scale-free, an abrupt behavior appears, with the order parameter remaining close to 0 up to a point at which it suddenly jumps to approximately 1. As the transition coupling is different for the forward and the backward diagrams, the transition is irreversible and displays a hysteresis loop.

An analytical approach is also possible for a specific star configuration, considered as an emblematic example of a heterogeneous network. The graph is composed of a central node (the hub h) and K peripheral nodes (the leaf nodes), each one labeled by the index i. By setting the frequency of the hub $\omega_h = K\omega$ and that of all the leaves $\omega_i = \omega$, the equations of motion can be written as

$$\dot{\vartheta}_i = \omega + \sigma \sin(\vartheta_h - \vartheta_i) \tag{5.156}$$

and

$$\dot{\vartheta}_h = K\omega + \sigma \sum_{i=1}^{K} \sin(\vartheta_i - \vartheta_h) . \tag{5.157}$$

Rewrite now Ψ as $\Psi(t) = \Psi(0) + \Omega t$, where Ω is the average frequency of the system. Using $\Psi(t)$ as a rotating frame, the equation for the evolution of the hub can be rewritten as

$$\dot{\vartheta}_h = (K - \Omega) + \sigma(K + 1) r \sin(\vartheta_h) . \tag{5.158}$$

The condition for stability of the hub locking solution $\dot{\vartheta}_h = 0$ allows one to obtain the *backward* critical coupling value $\sigma_c = (K - 1) / (K + 1)$. The order parameter at the critical point is $r_c = K / (K + 1)$, which is a positive value, confirming the existence of a discontinuity in the transition.

Starting from this result, several generalizations have been provided showing that explosive transitions to synchronization happen by no means only in heterogenous networks with correlations imposed ad hoc, but they are actually an emergent phenomenon in frequency-disassortative networks, and can occur also in homogeneous structures. For a general overview of explosive synchronization in complex networks, the reader is referred to an extensive monographic report on the topic (Boccaletti et al. 2016).

5.8 Synchronization in Temporal and Multilayer Networks

Recently, scientists have started modeling networked systems no longer as single static graphs, but as more complex structures, whose topologies can change in time (Holme and Saramäki 2012) or consist of multiple layers (Boccaletti et al. 2014). Here we discuss two main extensions of the Master Stability Function formalism, showing how it can be adapted to study synchronization in networks whose structure changes in time, and in networks containing multiple layers of interaction between the nodes.

5.8.1 Time-Varying Networks

Consider the case of a network of N oscillators whose structure of connections is allowed to change in time. To derive a Master Stability Function for this system, one can follow the same procedure as in Section 5.2.1, up to Equation 5.20, keeping in mind that now the matrices $\mathbf{V}$ and $\mathbf{D}$ depend on the time t. Then, define

$$\boldsymbol{\eta} \equiv \left(\mathbf{V}^{\mathsf{T}}(t) \otimes \mathbb{1}_m\right) \delta \mathbf{X}. \tag{5.159}$$

From this equation, it follows that

$$\dot{\boldsymbol{\eta}} = \left(\mathbf{V}^{\mathsf{T}}(t) \otimes \mathbb{1}_m\right) \delta \dot{\mathbf{X}} + \left(\frac{\mathrm{d}}{\mathrm{d}t}\mathbf{V}^{\mathsf{T}}(t) \otimes \mathbb{1}_m\right) \delta \mathbf{X}. \tag{5.160}$$

Then, using Equation 5.20, it is

$$\dot{\boldsymbol{\eta}} \approx \left(\mathbb{1}_N \otimes J\mathbf{f}(\mathbf{x_S}) - \sigma \mathbf{D}(t) \otimes J\mathbf{g}(\mathbf{x_S})\right)\boldsymbol{\eta} + \left(\frac{\mathrm{d}}{\mathrm{d}t}\mathbf{V}^{\mathsf{T}}(t) \otimes \mathbb{1}_m\right) \delta \mathbf{X}. \tag{5.161}$$

Using the relation

$$(\mathbf{M_1} \otimes \mathbf{M_2})^{-1} = \mathbf{M_1}^{-1} \otimes \mathbf{M_2}^{-1} \tag{5.162}$$

in Equation 5.159, we obtain

$$\delta \mathbf{X} = (\mathbf{V}(t) \otimes \mathbb{1}_m) \, \boldsymbol{\eta} \,, \tag{5.163}$$

where we used the fact that $\mathbf{V}$ is an orthogonal matrix. Then, Equation 5.161 becomes

$$\dot{\boldsymbol{\eta}} \approx (\mathbb{1}_N \otimes J\mathbf{f}(\mathbf{x}_S) - \sigma \mathbf{D}(t) \otimes J\mathbf{g}(\mathbf{x}_S)) \, \boldsymbol{\eta}$$
$$+ \left(\frac{d}{dt} \mathbf{V}^{\mathrm{T}}(t) \otimes \mathbb{1}_m \right) (\mathbf{V}(t) \otimes \mathbb{1}_m) \, \boldsymbol{\eta} \tag{5.164}$$
$$= (\mathbb{1}_N \otimes J\mathbf{f}(\mathbf{x}_S) - \sigma \mathbf{D}(t) \otimes J\mathbf{g}(\mathbf{x}_S)) \, \boldsymbol{\eta} + \left[\left(\frac{d}{dt} \mathbf{V}^{\mathrm{T}}(t) \right) \mathbf{V}(t) \otimes \mathbb{1}_m \right] \boldsymbol{\eta}.$$

Unlike what happens in the static case, the equation above does not decompose the evolution of $\mathbf{X}$ into decoupled eigenmodes, due to the second term in the right-hand side. This is again more evident if one rewrites the equation as a system:

$$\dot{\boldsymbol{\eta}}_{\mathbf{j}} \approx \left(J\mathbf{f}(\mathbf{x}_S) - \sigma \lambda_j(t) J\mathbf{g}(\mathbf{x}_S) \right) \boldsymbol{\eta}_{\mathbf{j}} + \sum_{i=1}^{N} \left[\left(\frac{d}{dt} \mathbf{v}_{\mathbf{j}}(t) \right) \cdot \mathbf{v}_{\mathbf{i}}(t) \right] \boldsymbol{\eta}_{\mathbf{i}} \,, \tag{5.165}$$

where $\mathbf{v}_i$ is the ith eigenvector of the graph Laplacian. Thus, to describe the evolution of the global synchronization error in time-varying networks, it is in general necessary to estimate the sum in the right-hand side of the equation above.

An exception to this general necessity happens when the structure of the network changes in a way that ensures that the graph Laplacians commute with each other at all times. In fact, if $[\mathbf{L}(t_1), \mathbf{L}(t_2)] = 0$ for all choices of t_1 and t_2, then there exists a common basis of eigenvectors that diagonalizes $\mathbf{L}$ at all times. In this case, the Laplacian eigenvectors are constant, their derivatives vanish, and Equation 5.165 reduces to Equation 5.22 (Boccaletti et al. 2006b).

If, instead, the evolution of the network topology is unconstrained, and the Laplacians do not necessarily commute, then one must consider the mixing term. Note that the particular structure of Equation 5.165 makes it suitable also for the study of a type of inverse problem, namely the construction of a suitable structural evolution between specified topologies. This introduces a series of constraints that must be satisfied. The most important is that the time derivatives in Equation 5.165 must exist at all times. This means that the evolution cannot include discontinuities, which would cause the derivatives of the eigenvectors to be not defined at some point in time.

This problem has been solved by describing a procedure to map the eigenvector matrices corresponding to the different structures via a proper rotation (del Genio et al. 2015). This map can then be used to reconstruct the actual structural evolution and compute the mixing term. The core of the technique is an iterative operator decomposition into invariant orthogonal subspaces. We omit the details here, referring the interested reader to the original publication. However, we note that the formalism developed in (del Genio et al. 2015) is slightly different from the one presented here. In fact, the main equation in that work reads

$$\dot{\eta}_j \approx \left(J\mathbf{f}\left(\mathbf{x}_S\right) - \sigma\lambda_j\left(t\right) J\mathbf{g}\left(\mathbf{x}_S\right)\right)\eta_j - \sum_{i=1}^{N}\left(\mathbf{v}_j\left(t\right)\cdot\frac{\mathrm{d}}{\mathrm{d}t}\mathbf{v}_i\left(t\right)\right)\eta_i \,. \tag{5.166}$$

Equations 5.165 and 5.166 are actually equivalent, as can easily be verified by differentiating Equation 5.163 and substituting it into Equation 5.15. Alternatively, one can obtain Equation 5.166 from Equation 5.165 using

$$\left(\frac{\mathrm{d}}{\mathrm{d}t}\mathbf{V}^{\mathrm{T}}\left(t\right)\right)\mathbf{V}\left(t\right) = -\mathbf{V}^{\mathrm{T}}\left(t\right)\frac{\mathrm{d}}{\mathrm{d}t}\mathbf{V}\left(t\right) \,, \tag{5.167}$$

which is a direct consequence of the orthogonality of $\mathbf{V}$.

5.8.2 Synchronization in Multilayer Networks

An extension of the MSF is possible also in the case of multilayer networks. The case we consider here is that of a network in which the nodes interact over more than one channel (del Genio et al. 2016). Thus, the different layers represent different types of interaction, while each node in each layer corresponds to the very same system element. In a network with M layers and N nodes, the evolution is described by

$$\dot{\mathbf{x}}_i = \mathbf{f}\left(\mathbf{x}_i\right) - \sum_{\alpha=1}^{M}\sigma_\alpha\sum_{j=1}^{N}L_{i,j}^{(\alpha)}\mathbf{g}_\alpha\left(\mathbf{x}_j\right) \,, \tag{5.168}$$

where each layer α has its own interaction strength σ_α, Laplacian $\mathbf{L}^{(\alpha)}$ and output function $\mathbf{g}_\alpha$.

To derive an MSF for this case, one can follow the same procedure as in Section 5.2.1 up to Equation 5.18, replacing the matrix of Laplacian eigenvectors $\mathbf{V}$ with the matrix of Laplacian eigenvectors of the first layer $\mathbf{V}^{(1)}$. Then, for the multilayer case, Equation 5.18 takes the form

$$\left(\mathbf{V}^{(1)^{\mathrm{T}}}\otimes\mathbb{1}_m\right)\delta\dot{\mathbf{X}}\approx\left[\mathbf{V}^{(1)^{\mathrm{T}}}\otimes J\mathbf{f}\left(\mathbf{x}_S\right) - \left(\sigma_1\mathbf{D}^{(1)}\mathbf{V}^{(1)^{\mathrm{T}}}\right)\otimes J\mathbf{g}_1\left(\mathbf{x}_S\right)\right]\delta\mathbf{X}$$

$$- \sum_{\alpha=2}^{M}\sigma_\alpha\left(\mathbf{V}^{(1)^{\mathrm{T}}}\mathbf{L}^{(\alpha)}\right)\otimes J\mathbf{g}_\alpha\left(\mathbf{x}_S\right)\delta\mathbf{X} \,, \tag{5.169}$$

where $\mathbf{D}^{(1)}$ is the diagonal matrix of the Laplacian eigenvalues of the first layer, and we have separated the contribution of the first layer from that of the other layers.

The equation above can be manipulated to into a more useful form. In the right-hand side, multiply the first occurrence of $\mathbf{V}^{(1)^\mathrm{T}}$ on the left by $\mathbb{1}_N$, and multiply $\mathbf{f}$ and $\mathbf{g_1}$ on the right by $\mathbb{1}_m$, to get

$$\left(\mathbf{V}^{(1)^\mathrm{T}} \otimes \mathbb{1}_m\right) \delta\dot{\mathbf{X}} \approx \left[\mathbb{1}_N \mathbf{V}^{(1)^\mathrm{T}} \otimes J\mathbf{f}\left(\mathbf{x_S}\right) \mathbb{1}_m \left(\sigma_1 \mathbf{D}^{(1)}\mathbf{V}^{(1)^\mathrm{T}}\right) \otimes J\mathbf{g_1}\left(\mathbf{x_S}\right) \mathbb{1}_m\right] \delta\mathbf{X}$$
$$- \sum_{\alpha=2}^{M} \sigma_\alpha \left(\mathbf{V}^{(1)^\mathrm{T}}\mathbf{L}^{(\alpha)}\right) \otimes J\mathbf{g_\alpha}\left(\mathbf{x_S}\right) \delta\mathbf{X}. \tag{5.170}$$

Using Equation 5.17, this becomes

$$\left(\mathbf{V}^{(1)^\mathrm{T}} \otimes \mathbb{1}_m\right) \delta\dot{\mathbf{X}} \approx \left[\left(\mathbb{1}_N \otimes J\mathbf{f}\left(\mathbf{x_S}\right)\right)\left(\mathbf{V}^{(1)^\mathrm{T}} \otimes \mathbb{1}_m\right)\right.$$
$$\left. - \left(\sigma_1 \mathbf{D}^{(1)} \otimes J\mathbf{g_1}\left(\mathbf{x_S}\right)\right)\left(\mathbf{V}^{(1)^\mathrm{T}} \otimes \mathbb{1}_m\right)\right]\delta\mathbf{X}$$
$$- \sum_{\alpha=2}^{M} \sigma_\alpha \left(\mathbf{V}^{(1)^\mathrm{T}}\mathbf{L}^{(\alpha)}\right) \otimes J\mathbf{g_\alpha}\left(\mathbf{x_S}\right) \delta\mathbf{X} \tag{5.171}$$
$$= \left(\mathbb{1}_N \otimes J\mathbf{f}\left(\mathbf{x_S}\right) - \sigma_1 \mathbf{D}^{(1)} \otimes J\mathbf{g_1}\left(\mathbf{x_S}\right)\right) \cdot \left(\mathbf{V}^{(1)^\mathrm{T}} \otimes \mathbb{1}_m\right) \delta\mathbf{X}$$
$$- \sum_{\alpha=2}^{M} \sigma_\alpha \left(\mathbf{V}^{(1)^\mathrm{T}}\mathbf{L}^{(\alpha)}\right) \otimes J\mathbf{g_\alpha}\left(\mathbf{x_S}\right) \delta\mathbf{X}.$$

Equation 5.162, together with the orthogonality of the $\mathbf{V}^{(\alpha)}$ matrices, implies that the Nm-dimensional identity matrix can be expressed as $\left(\mathbf{V}^{(1)} \otimes \mathbb{1}_m\right)$ $\left(\mathbf{V}^{(1)^\mathrm{T}} \otimes \mathbb{1}_m\right)$. Then, multiplying the last $\delta\mathbf{X}$ on the right-hand side of Equation 5.171 on the left by this expression, one gets

$$\left(\mathbf{V}^{(1)^\mathrm{T}} \otimes \mathbb{1}_m\right) \delta\dot{\mathbf{X}} \approx \left(\mathbb{1}_N \otimes J\mathbf{f}\left(\mathbf{x_S}\right) - \sigma_1 \mathbf{D}^{(1)} \otimes J\mathbf{g_1}\left(\mathbf{x_S}\right)\right)\left(\mathbf{V}^{(1)^\mathrm{T}} \otimes \mathbb{1}_m\right) \delta\mathbf{X}$$
$$- \sum_{\alpha=2}^{M} \sigma_\alpha \left(\mathbf{V}^{(1)^\mathrm{T}}\mathbf{L}^{(\alpha)}\right) \otimes J\mathbf{g_\alpha}\left(\mathbf{x_S}\right) \left(\mathbf{V}^{(1)} \otimes \mathbb{1}_m\right)\left(\mathbf{V}^{(1)^\mathrm{T}} \otimes \mathbb{1}_m\right) \delta\mathbf{X}, \tag{5.172}$$

which, with the definition

$$\boldsymbol{\eta} \equiv \left(\mathbf{V}^{(1)^\mathrm{T}} \otimes \mathbb{1}_m\right) \delta\mathbf{X}, \tag{5.173}$$

becomes

$$\dot{\boldsymbol{\eta}} \approx \left(\mathbb{1}_N \otimes J\mathbf{f}\left(\mathbf{x_S}\right) - \sigma_1 \mathbf{D}^{(1)} \otimes J\mathbf{g_1}\left(\mathbf{x_S}\right)\right) \boldsymbol{\eta}$$
$$- \sum_{\alpha=2}^{M} \sigma_\alpha \left(\mathbf{V}^{(1)^\mathrm{T}}\mathbf{L}^{(\alpha)}\right) \otimes J\mathbf{g_\alpha}\left(\mathbf{x_S}\right) \left(\mathbf{V}^{(1)} \otimes \mathbb{1}_m\right) \boldsymbol{\eta}. \tag{5.174}$$

As the $\mathbf{V}^{(\alpha)}$ matrices are orthogonal, the N-dimensional identity matrix can be written as $\mathbf{V}^{(\alpha)}\mathbf{V}^{(\alpha)T}$. Insert this expression in the equation above immediately before $\mathbf{L}^{(\alpha)}$ to get

$$\dot{\eta} \approx \left(\mathbb{1}_N \otimes J\mathbf{f}\left(\mathbf{x_S}\right) - \sigma_1 \mathbf{D}^{(1)} \otimes J\mathbf{g_1}\left(\mathbf{x_S}\right)\right)\eta$$

$$-\sum_{\alpha=2}^{M}\sigma_\alpha\left(\mathbf{V}^{(1)T}\mathbf{V}^{(\alpha)}\mathbf{V}^{(\alpha)T}\mathbf{L}^{(\alpha)}\right)\otimes J\mathbf{g}_\alpha\left(\mathbf{x_S}\right)\left(\mathbf{V}^{(1)}\otimes\mathbb{1}_m\right)\eta$$

$$=\left(\mathbb{1}_N\otimes J\mathbf{f}\left(\mathbf{x_S}\right) - \sigma_1\mathbf{D}^{(1)}\otimes J\mathbf{g_1}\left(\mathbf{x_S}\right)\right)\eta$$

$$-\sum_{\alpha=2}^{M}\sigma_\alpha\left(\mathbf{V}^{(1)T}\mathbf{V}^{(\alpha)}\mathbf{D}^{(\alpha)}\mathbf{V}^{(\alpha)T}\right)\otimes J\mathbf{g}_\alpha\left(\mathbf{x_S}\right)\left(\mathbf{V}^{(1)}\otimes\mathbb{1}_m\right)\eta \qquad (5.175)$$

$$=\left(\mathbb{1}_N\otimes J\mathbf{f}\left(\mathbf{x_S}\right) - \sigma_1\mathbf{D}^{(1)}\otimes J\mathbf{g_1}\left(\mathbf{x_S}\right)\right)\eta$$

$$-\sum_{\alpha=2}^{M}\sigma_\alpha\left(\mathbf{V}^{(1)T}\mathbf{V}^{(\alpha)}\mathbf{D}^{(\alpha)}\mathbf{V}^{(\alpha)T}\mathbf{V}^{(1)}\right)\otimes J\mathbf{g}_\alpha\left(\mathbf{x_S}\right)\eta,$$

where, in the last equality, we used once more Equation 5.17.

It is now convenient to define a matrix $\mathbf{\Gamma}^{(\alpha)} \equiv \mathbf{V}^{(\alpha)T}\mathbf{V}^{(1)}$. Then, the equation for $\dot{\eta}$ becomes

$$\dot{\eta} \approx \left(\mathbb{1}_N \otimes J\mathbf{f}\left(\mathbf{x_S}\right) - \sigma_1\mathbf{D}^{(1)}\otimes J\mathbf{g_1}\left(\mathbf{x_S}\right)\right)\eta$$

$$-\sum_{\alpha=2}^{M}\sigma_\alpha\left(\mathbf{\Gamma}^{(\alpha)T}\mathbf{D}^{(\alpha)}\mathbf{\Gamma}^{(\alpha)}\right)\otimes J\mathbf{g}_\alpha\left(\mathbf{x_S}\right)\eta. \qquad (5.176)$$

This equation consists of a purely variational part and a mixing term. The variational part consists of a block-diagonal matrix that does not mix the components of η. In the mixing part, instead, each component of η contributes to the others. As before, this effect is more evident when the equation is cast explicitly into the form of a system. To do so, note that the sum over α on the right-hand side defines an $Nm \times Nm$ block matrix. The mixing part contribution to the jth component of η is given by the product of the jth row of this block matrix by the vector η. But by definition of the Kronecker product, the kth block of the jth row of the block matrix is the $m \times m$ matrix $J\mathbf{g}_\alpha\left(\mathbf{x_S}\right)$ multiplied by the value of the (j, k) element of $\mathbf{\Gamma}^{(\alpha)T}\mathbf{D}^{(\alpha)}\mathbf{\Gamma}^{(\alpha)}$. In turn, this is

$$\left(\mathbf{\Gamma}^{(\alpha)T}\mathbf{D}^{(\alpha)}\mathbf{\Gamma}^{(\alpha)}\right)_{jk} = \sum_{r=1}^{N}\mathbf{\Gamma}^{(\alpha)T}_{jr}\lambda_r^{(\alpha)}\mathbf{\Gamma}^{(\alpha)}_{rk}. \qquad (5.177)$$

Using this expression, and summing over all the components η_k, one can rewrite Equation 5.176 as

$$\dot{\eta}_j \approx \left(J\mathbf{f}\left(\mathbf{x_S}\right) - \sigma_1 \lambda_j^{(1)} J\mathbf{g_1}\left(\mathbf{x_S}\right) \right) \eta_j$$

$$- \sum_{\alpha=2}^{M} \sigma_\alpha \sum_{k=2}^{N} \sum_{r=2}^{N} \lambda_r^{(\alpha)} \Gamma_{rj}^{(\alpha)} \Gamma_{rk}^{(\alpha)} J\mathbf{g}_\alpha\left(\mathbf{x_S}\right) \eta_k , \tag{5.178}$$

where the sums start from 2 because the first eigenvalue is always 0, and all the elements of the first row and column of $\mathbf{\Gamma}^{(\alpha)}$ are 0, except $\Gamma_{1,1}^{(\alpha)}$. Notice that the matrices $\mathbf{\Gamma}^{(\alpha)}$ measure the degree of alignment of the Laplacian eigenvectors of layer α with those of layer 1.

This makes it easy to see that, also in this case, if the Laplacians of all layers commute, one recovers a purely variational form. In fact, if the Laplacians commute, they can be simultaneously diagonalized by a common basis of eigenvectors. Then, all the $\mathbf{\Gamma}^{(\alpha)}$ are just the identity matrix, and Equation 5.178 becomes

$$\dot{\eta}_j \approx \left(J\mathbf{f}\left(\mathbf{x_S}\right) - \sigma_1 \lambda_j^{(1)} J\mathbf{g_1}\left(\mathbf{x_S}\right) \right) \eta_j$$

$$- \sum_{\alpha=2}^{M} \sigma_\alpha \sum_{k=2}^{N} \sum_{r=2}^{N} \lambda_r^{(\alpha)} \delta_{r,j} \delta_{r,k} J\mathbf{g}_\alpha\left(\mathbf{x_S}\right) \eta_k$$

$$= \left(J\mathbf{f}\left(\mathbf{x_S}\right) - \sum_{\alpha=1}^{N} \sigma_\alpha \lambda_j^{(\alpha)} J\mathbf{g}_\alpha\left(\mathbf{x_S}\right) \right) \eta_j . \tag{5.179}$$

5.8.3 Application of the Master Stability Function to Multistable Dynamical Systems

The Master Stability Function (MSF) can also be applied to systems with more complex dynamical behavior, such as multistability. However, in this case, one should study the MSF of each attractor separately.

Figure 5.10(a) shows the MSFs corresponding to both the small-amplitude attractor S (continuous line) and large-amplitude attractor L (dashed line) for the Rössler-like system with diffusive coupling through the x variable as described by Equation 3.13. According to the classification discussed in Section 5.2.2, this system belongs to class III, as the region of stable synchronization is bounded between two MSF zeros. Interestingly, Figure 5.10 also shows that the stability region of L is contained within that of S. The interplay between the two stability regions given by the MSFs explains how the synchronized state is maintained (or lost) in one of the three possible scenarios:

(i) Neither S nor L can be synchronized, i.e., the MSF is positive for both S and L.

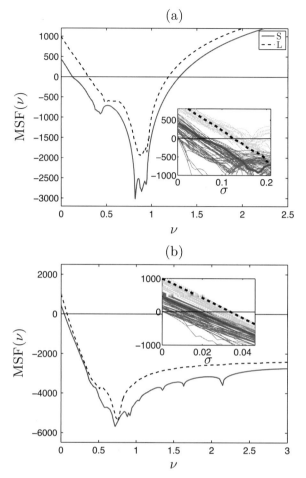

Figure 5.10 Master Stability Function for coexisting attractors for two types of coupling, and variability with respect to uncertainties in parameters. Panel (a) refers to oscillators coupled through the x variable (class III system), and panel (b) to oscillators coupled through the y variable (class II system). The insets show the variability of the first zero crossing as the parameters are affected by uncertainties. In both panels, the continuous line corresponds to the S attractor and the dashed line to the L attractor.

(ii) S can be synchronized, but L can not, i.e., the MSF is negative for S, but positive for L.

(iii) Both S and L can be synchronized, i.e., the MSF is negative for both attractors.

Case (ii) only guarantees the synchronization stability in a regime in which the system ends up in an attractor indistinguishable from S. Case (iii), instead, means that synchronization is stable no matter how complex the dynamics may be, even in the presence of intermittency. Similar arguments can be used in systems with an arbitrarily large number of attractors, provided the stability regions of the different attractors are not disjoint.

Figure 5.10(b) shows the MSFs corresponding to the two coexisting attractors when the oscillators are coupled through the y variable. In this case the system belongs to class II, i.e., the stability region starts at a given v and then extends indefinitely to the right. Again, for this particular system the stability region of L is contained in that of S. Choosing a high enough σ, or suitably changing the topology for a given σ, makes all the eigenmodes lie in the region to the right of the MSF zero for L, guaranteeing the stability of synchronous dynamics on any attractor, even in the presence of intermittency.

The existence of interwoven basins of attraction hinders the prediction of the asymptotic behavior of the system in the presence of noise or parameter uncertainties. The methodology described here is of special interest for a stability analysis of synchronization in multistable systems in the presence of intermittency, since any monostable approach to synchronization is bound to fail in that regime. Knowing the minimum coupling strength or the topological modifications required to maintain complete synchronization under any possible attractor dynamics is especially useful in this scenario.

5.9 Single-Oscillator Experiments

The practical realization of a complex network with a very large number of chaotic oscillators may be a step towards the creation of artificial intelligence and a better understanding of the functioning of the brain. However, traditional methods for constructing complex networks pose serious technical difficulties that have restricted the experimental studies to relatively small systems.

An approach based on recording the output of a single system in a memory to study synchronization between two unidirectionally coupled identical chaotic units was introduced by Pyragas (1993). Its usefulness was experimentally demonstrated with a chaotic electronic circuit by Kittel and collaborators (1994). Later, this idea was extended to a network of coupled oscillators (Pisarchik et al. 2009). Below, we show how a single oscillator can be used to study synchronization in a large complex network of unidirectionally interacting units with a given topology.

5.9.1 Experimental Setup

The experimental arrangement is shown in Figure 5.11. It consists of a nonlinear oscillator, an analog–digital converter (ADC), a personal computer (PC), a digital–analog converter (DAC), and an operational amplifier (OA).

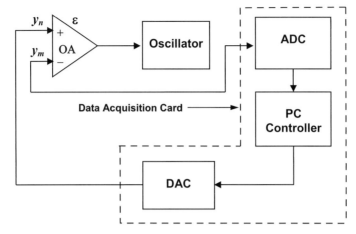

Figure 5.11 Experimental setup for studying complex network synchronization. ADC and DAC are analog–digital and digital–analog converters, PC is a personal computer, and OA is an operational amplifier (reprinted with permission from Pisarchik et al. 2009).

A chaotic waveform generated by the oscillator passes through the ADC and is stored in a computer. Then, after passing through a DAC and an OA, the digital signal from the computer is fed back into the same oscillator to provide the coupling between nodes, which form a network with a predefined configuration.

With this setup, the oscillator first acts as node 1. Then, its analog output passes through the ADC and is stored in the PC, where it is analyzed by an appropriate program. After attenuation, it goes through the DAC, and, multiplied by the coupling strength between node 1 and node 2, it enters the oscillator again. The oscillator then acts as node 2, and this process is repeated until the oscillator has acted as each node in the network.

If a node has several links, the computer adjusts the input signal according to the connectivity matrix. In principle, this experimental setup simulates a network of identical dynamical systems. However, since the experimental noise can act differently on the same oscillators at different times, the systems are more precisely described as being almost identical. Since the waveforms are stored in the computer and then injected back into the analog system as an external driving signal, and the oscillator is self-oscillating, the node needs time to be synchronized. Therefore, the stored time series must be much larger than the transient time to achieve pairwise synchronization.

5.9.2 General Equation

With the usual notation, the general equation of motion of a network built by N unidirectionally coupled identical oscillators can be written as

$$\dot{\mathbf{x}}_{\mathbf{i}} = \mathbf{f}(\mathbf{x}_{\mathbf{i}}) - \varepsilon \sum_{j=1}^{N} L_{ij} \mathbf{g}(\mathbf{x}_{\mathbf{j}}) . \tag{5.180}$$

Here, $\varepsilon \in [0, 1]$ is the coupling strength, which is assumed constant for the sake of simplicity.

To demonstrate the applicability of the experimental approach, let us consider two types of coupling schemes, one in which all the *incoming* links to the same node have the same weight and unitary sum, and one in which all the *outgoing* links from the same node have the same weight and unitary sum.

In the former case, which we call the *informational scheme*, the weight of each link entering a node i is the inverse of the in-degree of that node:

$$W_{ij} = \begin{cases} 0 & \text{if node } j \text{ does not link to node } i \\ \frac{1}{k_i^{(in)}} & \text{otherwise.} \end{cases} \tag{5.181}$$

This coupling scheme can simulate, in general, any network where each node receives a fixed amount of information regardless of the number of sources. This way, the weight sum of all the links entering a node i is $\sum_{j=1}^{N} W_{ij} = k_i^{(in)} \left(1/k_i^{(in)}\right) = 1$.

In the latter case, which we call the *commodity scheme*, the weight of each link exiting a node j is the inverse of the out-degree of that node:

$$W_{ij} = \begin{cases} 0 & \text{if node } j \text{ does not link to node } i \\ \frac{1}{k_j^{(out)}} & \text{otherwise.} \end{cases} \tag{5.182}$$

Thus, the weight sum of all the links exiting a node j is $\sum_{i=1}^{N} W_{ij} = k_j^{(out)} \left(1/k_j^{(out)}\right) = 1$. Such a scheme can describe commodity exchanges between factories or transport companies, or, in general, any network in which units generate products to be distributed evenly among their connections.

For the simplest case, where the oscillators are coupled by only one variable y, the last term in Equation 5.180 can be replaced by $\frac{\varepsilon}{k_i^{(in)}} \sum_{j=1}^{N}(y_j - y_i)$ and $\varepsilon \sum_{j=1}^{N} \frac{1}{k_j^{(out)}}(y_j - y_i)$ for the informational and commodity schemes, respectively. Note that these networks are not node-balanced, because the input and output weight sums are not equal. Nevertheless, they are very common in practice and their correct operation depends on how well they are synchronized.

We are now interested in answering the following questions:

(i) What are the differences between the synchronization properties of these two network schemes?

(ii) For each network type, what is the optimal coupling to obtain the best synchronization between any two chosen nodes?

(iii) Does the optimal coupling depend on network configuration?

In order to provide some answers, we need to show how the synchronization between every pair of nodes depends on the coupling strength in both network types. For simplicity, we consider a network formed by piecewise linear Rössler-like oscillators, already described in Chapter 3.

5.9.3 Network of Piecewise Rössler Oscillators

The dynamics of the ith oscillator are given by

$$
\begin{aligned}
\frac{d}{dt} x_i &= -\alpha x_i - z_i - \beta y_i , \\
\frac{d}{dt} y_i &= x_i + \gamma y_i , \\
\frac{d}{dt} z_i &= g(x_i) - z_i ,
\end{aligned}
\tag{5.183}
$$

where x_i, y_i, and z_i are the oscillator state variables for node i, and g is the piecewise linear function

$$
g(x_i) = \begin{cases} 0 & \forall x_i \leqslant 3 \\ \mu(x_i - 3) & \forall x_i > 3 . \end{cases}
\tag{5.184}
$$

For the choice of the parameters $\alpha = 0.05$, $\beta = 0.5$, $\gamma = 0.266$, and $\mu = 15$, these oscillators generate chaotic waveforms. Also, we let the coupling between systems be on the state variable y.

For a better illustration, we first consider a small network formed by only six identical chaotic oscillators. Figure 5.12 shows the network configuration with link weights for the informational and commodity connections (panels (a) and (b), respectively). The corresponding Laplacians are

$$
\mathbf{L}^{\text{inf}} = \begin{pmatrix}
0 & 0 & 0 & 0 & 0 & 0 \\
-1 & 1 & 0 & 0 & 0 & 0 \\
0 & -1 & 1 & 0 & 0 & 0 \\
-\frac{1}{3} & -\frac{1}{3} & -\frac{1}{3} & 1 & 0 & 0 \\
0 & -\frac{1}{2} & 0 & -\frac{1}{2} & 1 & 0 \\
-\frac{1}{3} & 0 & -\frac{1}{3} & 0 & -\frac{1}{3} & 1
\end{pmatrix}
\tag{5.185}
$$

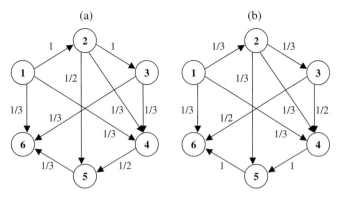

Figure 5.12 (a) Informational and (b) commodity coupling schemes for a six-node network (reprinted with permission from Pisarchik et al. 2009).

and

$$
\mathbf{L}^{\text{com}} =
\begin{pmatrix}
0 & 0 & 0 & 0 & 0 & 0 \\
-\frac{1}{3} & \frac{1}{3} & 0 & 0 & 0 & 0 \\
0 & -\frac{1}{3} & \frac{1}{3} & 0 & 0 & 0 \\
-\frac{1}{3} & -\frac{1}{3} & -\frac{1}{2} & \frac{7}{6} & 0 & 0 \\
0 & -\frac{1}{3} & 0 & -1 & \frac{4}{3} & 0 \\
-\frac{1}{3} & 0 & -\frac{1}{2} & 0 & -1 & \frac{11}{6}
\end{pmatrix}.
\tag{5.186}
$$

5.9.4 Synchronization Error

Synchronization of complex networks can be quantitatively described by the average synchronization error $\langle e \rangle$ between any pair of oscillators in the network.

Figure 5.13 shows the error $\langle e \rangle = \langle y_i - y_1 \rangle$ computed for all nodes with respect to node 1 as a function of ε for the example networks in both schemes. The relations are not monotonic and their behaviors are determined by the occurrence of different types of synchronization. In particular, the maxima are due to either antiphase synchronization or intermittent antiphase synchronization, whereas the minima correspond to phase synchronization.

Numerical simulations of the system yield good agreement with the experimental results. The main advantages of the experimental approach over simulations are that typically it is much faster to generate long chaotic waveforms with an analog system than to obtain them numerically from complex nonlinear models, and that the experiments can be performed with devices whose model and parameters are not exactly known, such as lasers. More precisely, it can be shown that the time difference between experimental and numerical methods scales with the numbers

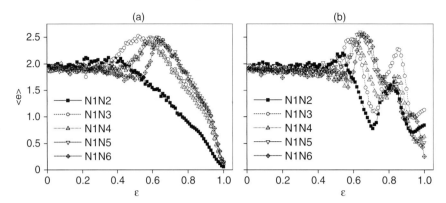

Figure 5.13 Average synchronization error between node 1 and other nodes for a network of six oscillators coupled with (a) informational and (b) commodity schemes, as a function of coupling strength. The nodes between which the synchronization error is measured are indicated in the bottom-left corner (reprinted with permission from Pisarchik et al. 2009).

of nodes and links, making the experimental approach more practical for larger networks.

Experiments with large networks reveal the main difference between informational and commodity schemes: for the informational type, the optimal coupling, corresponding to the minimum of $\langle e \rangle$, is always $\varepsilon_{opt} = 1$ for any pair of oscillators regardless of the network configuration, while for the commodity type ε_{opt} is different for each pair of nodes and does depend on the network topology. Note that the experimental setup is time-independent; for example, if node 1 is connected with nodes 2 and 1000, the same waveform is used for driving both nodes, and both may simultaneously synchronize with node 1, after a transient time that depends on the initial conditions for each oscillator.

References

Abarbanel, H. D. I., Rulkov, N. F., and Sushchik, M. M. 1996. Generalized synchronization of chaos: The auxiliary system approach. *Phys. Rev. E*, **53**(5), 4528–4535.

Abrams, D. M. and Strogatz, S. H. 2004. Chimera states for coupled oscillators. *Phys. Rev. Lett.*, **93**, 174102.

Abrams, D. M. and Strogatz, S. H. 2006. Chimera states in a ring of nonlocally coupled oscillators. *Int. J. Bif. Chaos*, **16**, 21–37.

Abrams, D. M., Mirollo, R., Strogatz, S. H., and Wiley, D. A. 2008. Solvable model for chimera states of coupled oscillators. *Phys. Rev. Lett.*, **101**, 084103.

Acebrón, J. A., Bonilla, L. L., Pérez Vicente, C. J., Ritort, F., and Spigler, R. 2005. The Kuramoto model: A simple paradigm for synchronization phenomena. *Rev. Mod. Phys.*, **77**, 137–185.

Adler, R. 1946. A study of locking phenomena in oscillators. *Proceedings of the I.R.E. and Waves and Electrons*, **34**, 351–357.

Afraimovich, V. S., Verichev, N. N., and Rabinovich, M. I. 1986. Stochastic synchronization of oscillation in dissipative systems. *Radiophys. Quant. Electron.*, **29**, 795–803.

Aizawa, Y. 1976. Synergetic approach to the phenomena of mode-locking in nonlinear systems. *Progr. Theor. Phys.*, **56**, 703–716.

Albert, R. and Barabási, A.-L. 2002. Statistical mechanics of complex networks. *Rev. Mod. Phys.*, **74**, 47–97.

Alexander, J. C., Yorke, J. A., You, Z., and Kan, I. 1992. Riddled basins. *Int. J. Bif. Chaos*, **2**, 795–813.

Amann, A. and Hooton, E. W. 2013. An odd-number limitation of extended time-delayed feedback control in autonomous systems. *Phil. Trans. R. Soc. A*, **371**, 20120463.

Amann, A., Schöll, E., and Just, W. 2007. Some basic remarks on eigenmode expansions of time-delaydynamics. *Physica A*, **373**, 191–202.

Amit, D. J. 1995. The Hebbian paradigm reintegrated: Local reverberations as internal representations. *Behav. Brain Sci.*, **18**, 617–657.

Andronov, A. A. and Witt, A. A. 1930. Towards mathematical theory of capture. *Arch. Electrotechnik (Berlin)*, **24**, 99–110.

Anishchenko, V. S., Vadivasova, T. E., Kopeikin, A. S., Kurths, J., and Strelkova, G. I. 2001. Effect of noise on the relaxation to an invariant probability measure of nonhyperbolic chaotic attractors. *Phys. Rev. Lett.*, **87**, 054101.

Arecchi, F. T., Meucci, R., Puccioni, G., and Tredicce, J. 1982. Experimental evidence of subharmonic bifurcations, multistability, and turbulence in a Q-switched gas laser. *Phys. Rev. Lett.*, **49**, 1217–1220.

Arecchi, F. T., Meucci, R., Allaria, E., Di Garbo, A., and Tsimring, L. S. 2002. Delayed self-synchronization in homoclinic chaos. *Phys. Rev. E*, **65**, 046237.

Argyris, A., Syvridis, D., Larger, L. et al. 2005. Chaos-based communications at high bit rates using commercial fibre-optic links. *Phys. Rev. Lett.*, **438**(7066), 343–346.

Arnéodo, A., Coullet, P., and Tresser, C. 1981. A possible new mechanism for the onset of turbulence. *Phys. Lett. A*, **81**, 197–201.

Arnold, V. I. 1974. *Mathematical Methods of Classical Mechanics*. Berlin: Springer-Verlag.

Aronson, D. G., Ermentrout, G. B., and Kopell, N. 1990. Amplitude response of coupled oscillators. *Physica D*, **41**(3), 403–449.

Ashwin, P. 2003. Synchronization from chaos. *Nature*, **422**, 384–385.

Ashwin, P. and Burylko, O. 2015. Weak chimeras in minimal networks of coupled phase oscillators. *Chaos*, **25**(1), 013106.

Astakhov, V., Shabunin, A., Uhm, W., and Kim, S. 2001. Multistability formation and synchronization loss in coupled Hénon maps: Two sides of the single bifurcational mechanism. *Phys. Rev. E*, **63**, 056212.

Atteneave, F. 1971. Multistability in perception. *Scientific American*, **225**, 63–71.

Baker, G. L. and Gollub, J. P. 1996. *Chaotic Dynamics: An Introduction*. Second edn. Cambridge: Cambridge University Press.

Baptista, M. S. 1998. Cryptography with chaos. *Phys. Lett. A*, **240**, 50–54.

Bar-Eli, K. 1985. On the stability of coupled chemical oscillators. *Physica D*, **14**, 242–252.

Barabási, A.-L. and Albert, R. 1999. Emergence of scaling in random networks. *Science*, **286**, 509–512.

Barahona, M. and Pecora, L. M. 2002. Synchronization in small-world systems. *Phys. Rev. Lett.*, **89**, 054101.

Basa, E. (ed.) 1990. *Chaos in Brain Function*. Springer Series in Brain Dynamics. Berlin: Springer-Verlag.

Bassler, K. E., del Genio, C. I., Erdös, P., Miklós, I., and Toroczkai, Z. 2015. Exact sampling of graphs with prescribed degree correlations. *New J. Phys.*, **17**, 083052.

Bellomo, N., Bianca, C., and Delitala, M. 2009. Complexity analysis and mathematical tools towards the modelling of living systems. *Phys. Life Revs.*, **6**, 144–175.

Belykh, I., de Lange, E., and Hasler, M. 2005. Synchronization of bursting neurons: What matters in the network topology. *Phys. Rev. Lett.*, **94**, 188101.

Bennet, M., Schatz, M. F., Rockwood, H., and K., Wiesenfeld. 2002. Huygens' clocks. *Proc. R. Soc. Lond. A*, **458**, 563–579.

Bertram, M., Beta, C., Pollmann, M. et al. 2003. Pattern formation on the edge of chaos: Experiments with CO oxidation on a Pt(110) surface under global delayed feedback. *Phys. Rev. E*, **67**, 036208.

Bhattacharya, J. and Petsche, H. 2001. Enhanced phase synchrony in the electroencephalograph γ band for musicians while listening to music. *Phys. Rev. E*, **64**, 012902.

Bi, G. Q. and Poo, M. M. 1998. Synaptic modifications in cultured hippocampal neurons: dependence on spike timing, synaptic strength, and postsynaptic cell type. *J. Neurosci.*, **18**, 10464.

Bi, H., Hu, X., Boccaletti, S. et al. 2016. Coexistence of quantized, time dependent, clusters in globally coupled oscillators. *Phys. Rev. Lett.*, **117**, 204101.

Blasius, B., and Stone, L. 2000. Chaos and phase synchronization in ecological systems. *Int. J. Bif. Chaos*, **10**, 2361–2380.

Blasius, B., Huppert, A., and Stone, L. 1999. Complex dynamics and phase synchronization in spatially extended ecological systems. *Nature*, **399**, 354–359.

Bob, P. 2007. Chaos, brain and divided consciousness. *Acta Univ. Carol. Med. Monogr.*, **153**, 9–80.

Boccaletti, S. 2008. *The Synchronized Dynamics of Complex Systems*. Ámsterdam: Elsevier.

Boccaletti, S., Pecora, L. M., and Pelaez, A. 2001. Unifying framework for synchronization of coupled dynamical systems. *Phys. Rev. E*, **63**, 066219.

Boccaletti, S., Kurths, J., Osipov, G., Valladares, D. L., and Zhou, C. S. 2002. The synchronization of chaotic systems. *Phys. Rep.*, **366**, 1–101.

Boccaletti, S., Latora, V., Moreno, Y., Chavez, M., and Hwanga, D.-U. 2006a. Complex networks: Structure and dynamics. *Phys. Rep.*, **424**, 175–308.

Boccaletti, S., Hwang, D.-U., Chavez, M. et al. 2006b. Synchronization in dynamical networks: Evolution along commutative graphs. *Phys. Rev. E*, **74**, 016102.

Boccaletti, S., Bianconi, G., Criado, R. et al. 2014. The structure and dynamics of multilayer networks. *Phys. Rep.*, **544**, 1–122.

Boccaletti, S., Almendral, J. A., Guan, S. et al. 2016. Explosive transitions in complex networks' structure and dynamics: percolation and synchronization. *Phys. Rep.*, **660**, 1–94.

Bragard, J., Vidal, G., Mancini, H., Mendoza, C., and Boccaletti, S. 2007. Chaos suppression through asymmetric coupling. *Chaos*, **17**, 043107.

Brown, R. and Rulkov, N. F. 1997a. Designing a coupling that guarantees synchronization between identical chaotic systems. *Phys. Rev. Lett.*, **78**, 4189–4192.

Brown, R. and Rulkov, N. F. 1997b. Synchronization of chaotic systems: Transverse stability of trajectories in invariant manifolds. *Chaos*, **7**(3), 395–413.

Brown, R., Rulkov, N. F., and Tracy, E. R. 1994. Modeling and synchronizing chaotic systems from time-series data. *Phys. Rev. E*, **49**, 3784–3800.

Brun, E., Derighetti, B., Meier, D., Holzner, R., and Ravani, M. 1985. Observation of order and chaos in a nuclear spin-flip laser. *J. Opt. Soc. Am. B*, **2**, 156–167.

Bruno, L., Tosin, A., Tricerri, P., and Venuti, F. 2011. Non-local first-order modelling of crowd dynamics: A multidimensional framework with applications. *Appl. Math. Mod.*, **35**(1), 426–445.

Buldú, J. M., Heil, T., Fisher, I., Torrent, M. C., and García-Ojaivo, J. 2006. Episodic synchronization via dynamic injection. *Phys. Rev. Lett.*, **96**, 024102.

Buzsáki, G. (ed.) 2006. *Rhythms of the Brain*. New York: Oxford University Press.

Buzsáki, G. and Chrobak, J. J. 1995. Temporal structure in spatially organized neuronal ensembles: a role for interneuronal networks. *Curr. Opin. Neurobiol.*, **5**(4), 504–510.

Camazine, S., Deneubourg, J.-L., Franks, N. R., Sneyd, J., Theraulaz, G., and Bonabeau, E. 2001. *Self-Organization in Biological Systems*. Princeton, NJ: Princeton University Press.

Canavier, C., Baxter, D., Clark, J., and Byrne, J. 1993. Nonlinear dynamics in a model neuron provide a novel mechanism for transient synaptic inputs to produce long-term alterations of postsynaptic activity. *J. Neurophysiol.*, **69**, 2252–2257.

Cao, H. and Sanjuan, M. A. F. 2009. A mechanism for elliptic-like bursting and synchronization of bursts in a map-based neuron network. *Cogn. Process.*, **10**, S23.

Carroll, T. L. 2001. Noise-resistant chaotic synchronization. *Phys. Rev. E*, **64**, 015201.

Castellano, C., Fortunato, S., and Loreto, V. 2009. Statistical physics of social dynamics. *Rev. Mod. Phys.*, **81**, 591–646.

Chavez, M., Hwang, D.-U., Amann, A., Hentschel, H. G. E., and Boccaletti, S. 2005. Synchronization is enhanced in weighted complex networks. *Phys. Rev. Lett.*, **94**, 218701.

Chen, J. Y., Wong, K. W., Zheng, H. Y., and Shuai, J. W. 2001. Intermittent phase synchronization of coupled spatiotemporal chaotic systems. *Phys. Rev. E*, **64**, 016212.

Chialvo, D. R. 1995. Generic excitable dynamics on a two-dimensional map. *Chaos, Solitons and Fractals*, **5**(3), 461–479.

Chiba, H. 2015. A proof of the Kuramoto conjecture for a bifurcation structure of the infinite-dimensional Kuramoto model. *Erg. Th. Dyn. Sys.*, **35**, 762–834.

Chiba, H. and Pazó, D. 2009. Stability of an $N/2$-dimensional invariant torus in the Kuramoto model at small coupling. *Physica D*, **238**, 1068–1081.

Chizhevsky, V. N., Corbalán, R., and Pisarchik, A. N. 1997. Attractor splitting induced by resonant perturbations. *Phys. Rev. E*, **56**, 1580–1584.

Cohen, R. and Havlin, S. 2003. Scale-free networks are ultrasmall. *Phys. Rev. Lett.*, **90**, 058701.

Copelli, M., Tragtenberg, M. H. R., and Kinouchi, O. 2004. Stability diagrams for bursting neurons modeled by three-variable maps. *Physica A*, **342**, 263–269.

Corron, N. J., Blakely, J. N., and Pethel, S. D. 2005. Lag and anticipating synchronization without time-delay coupling. *Chaos*, **15**, 023110.

Coscia, V. and Canavesio, C. 2008. First-order macroscopic modelling of human crowd dynamics. *Math. Models and Methods Appl. Sciences*, **18**(supp01), 1217–1247.

Cramer, A. O. J., Waldorp, L. J., van der Maas, H. L. J., et al. 2010. Comorbidity: A network perspective. *Behav. Brain Sci.*, **33**(2–3), 137–150.

Crank, J. 1980. *The Mathematics of Diffusion*. Oxford: Oxford University Press.

Crutchfield, J. P. and Kaneko, K. 1988. Are attractors relevant to turbulence? *Phys. Rev. Lett.*, **60**, 2715–2718.

Csicsvari, J., Hirase, H., Mamiya, A., and Buzsáki, G. 2000. Ensemble patterns of hippocampal CA3-CA1 neurons during sharp wave-associated population events. *Neuron*, **28**, 585–594.

Cuomo, K. M. and Oppenheim, A. V. 1993. Circuit implementation of synchronized chaos with applications to communications. *Phys. Rev. Lett.*, **71**, 65–68.

Doedel, E. J., Champneys, A. R., Fairgrieve, T. et al. 2007. *AUTO-07P: Continuation and bifurcation software for ordinary differential equations*. Tech. rept. Concordia University Montreal, Canada.

Dostalkova, I. and Spinka, M. 2010. When to go with the crowd: Modelling synchronization of all-or-nothing activity transitions in grouped animals. *J. Theor. Biol.*, **263**(4), 437–448.

Eckmann, J.-P. 1981. Roads to turbulence in dissipative dynamical systems. *Rev. Mod. Phys.*, **53**, 643–654.

Ekert, A. K. 1991. Quantum cryptography based on Bell's theorem. *Phys. Rev. Lett.*, **67**, 661–663.

Elson, R. C., Selverston, A. I., Huerta, R., Rulkov, N. F., Rabinovich, M. I., and Abarbanel, H. D. I. 1998. Synchronous behavior of two coupled biological neurons. *Phys. Rev. Lett.*, **81**(25), 5692–5695.

Elson, R. C., Selverston, A. I., Abarbanel, H. D. I., and Rabinovich, M. 2002. Inhibitory synchronization of bursting in biological neurons: Dependence on synaptic time constant. *J. Neurophysiol.*, **88**(10.1152/jn.00784.2001), 1166–1176.

Ermentrout, G. B. and Kopell, N. 1990. Oscillator death in systems of coupled neural oscillators. *SIAM J. Appl. Math.*, **50**, 125–146.

Ermentrout, G. B., Glass, L., and Oldeman, B. E. 2012. The shape of phase-resetting curves in oscillators with a saddle node on an invariant circle bifurcation. *Neural Comput.*, **24**, 3111–3125.

Fiedler, B., Flunkert, V., Georgi, M., Hövel, P., and Schöll, E. 2007. Refuting the odd number limitation of time-delayed feedback control. *Phys. Rev. Lett.*, **98**, 114101.

Fiedler, M. 1972. Bounds for eigenvalues of doubly stochastic matrices. *Linear Algebra Appl.*, **5**, 299–310.

Fiedler, M. 1973. Algebraic connectivity of graphs. *Czech. Math. J.*, **23**, 298–305.

Fisher, R. S., van Emde Boas, W., Blume, W., Elger, C., Genton, P., Lee, P., and Engel, J. 2005. Epileptic seizures and epilepsy: Definitions proposed by the International League Against Epilepsy (ILAE) and the International Bureau for Epilepsy (IBE). *Epilepsia*, **46**(4), 470–472.

Foss, J., Longtin, A., Mensour, B., and Milton, J. 1996. Multistability and delayed recurrent loops. *Phys. Rev. Lett.*, **76**, 708–711.

Franke, J. E. and Selgrade, J. F. 1976. Abstract ω-limit sets, chain recurrent sets, and basic sets for flows. *Proceedings of the American Mathematical Society*, **60**(1), 309–316.

Franoviĉ, I. and Miljkoviĉ, V. 2010. Power law behavior related to mutual synchronization of chemically coupled map neurons. *Eur. Phys. J. B*, **76**, 613–624.

Freud, S. 1922. *Group Psychology and the Analysis of the Ego*. New York: Boni and Liveright.

Fridrich, J. 1998. Symmetric ciphers based on two-dimensional chaotic maps. *Int. J. Bif. Chaos*, **8**(6), 1259–1284.

Fuchs, A. 2013. *Nonlinear Dynamics in Complex Systems*. Berlin: Springer-Verlag.

Fujisaka, H. and Yamada, T. 1983. Stability theory of synchronized motion in coupled-oscillator system. *Prog. Theor. Phys.*, **69**, 32–47.

Fujisaka, H., Kamifukumoto, H., and M., Inoue. 1983. Intermittency associated with the breakdown of the chaos symmetry. *Prog. Theor. Phys.*, **69**, 333–337.

Gabor, D. 1946. Theory of communication. *Proc. IEE London*, **93**, 429–457.

García-López, J. H., R. Jaimes-Reátegui, R., Pisarchik, A. N. et al. 2005. Novel communication scheme based on chaotic Rössler circuits. *J. Phys.: Conf. Ser.*, **23**, 175–184.

García-López, J. H., Jaimes-Reátegui, R., Chiu-Zarate, R. et al. 2008. Secure computer communication based on chaotic Rössler oscillators. *Open Electrical and Electronic Engineering Journal*, **2**, 41–44.

García-Vellisca, M. A., Pisarchik, A. N., and Jaimes-Reátegui, R. 2016. Experimental evidence of deterministic coherence resonance in coupled chaotic oscillators with frequency mismatch. *Phys. Rev. E*, **94**, 012218.

Gauthier, D. J. and Bienfang, J. C. 1996. Intermittent loss of synchronization in coupled chaotic oscillators: Toward a new criterion for high-quality synchronization. *Phys. Rev. Lett.*, **77**, 1751–1754.

del Genio, C. I., Gómez-Gardeñes, J., Bonamassa, I., and Boccaletti, S. 2016. Synchronization in networks with multiple interaction layers. *Sci. Adv.*, **2**(11), e1601679.

del Genio, C. I., Gross, T., and Bassler, K. E. 2011. All scale-free networks are sparse. *Phys. Rev. Lett.*, **107**, 178701.

del Genio, C. I., Kim, H., Toroczkai, Z., and Bassler, K. E. 2010. Efficient and exact sampling of simple graphs with given arbitrary degree sequence. *PLoS One*, **5**, e10012.

del Genio, C. I., Romance, M., Criado, R., and Boccaletti, S. 2015. Synchronization in dynamical networks with unconstrained structure switching. *Phys. Rev. E*, **92**, 062819.

Gladwell, M. 2000. *The Tipping Point: How Little Things Can Make a Big Difference*. Boston: Little, Brown and Company.

Goel, P. and Ermentrout, B. 2002. Synchrony, stability, and firing patterns in pulse-coupled oscillators. *Physica D*, **163**, 191–216.

Gómez-Gardeñes, J., Gómez, S., Arenas, A., and Moreno, Y. 2011. Explosive synchronization transitions in scale-free networks. *Phys. Rev. Lett.*, **106**, 128701.

Gong, P., Nikolaev, A. R., and van Leeuwen, C. 2007. Intermittent dynamics underlying the intrinsic fluctuations of the collective synchronization patterns in electrocortical activity. *Phys. Rev. E*, **76**, 011904.

González-Miranda, J. M. 2004. *Synchronization and Control of Chaos: An Introduction for Scientists and Enginners*. London: Imperial College Press.

Grebogi, C., Ott, E., and Yorke, J. A. 1983. Fractal basin boundaries, long-lived chaotic transients, and unstable–unstable pair bifurcation. *Phys. Rev. Lett.*, **50**(13), 935–938.

Grebogi, C., Ott, E., and Yorke, J. A. 1988. Unstable periodic orbits and the dimensions of multifractal chaotic attractors. *Phys. Rev. A*, **37**, 1711–1724.

Guan, S., Lai, C. H., and Wei, G. W. 2005. Bistable chaos without symmetry in generalized synchronization. *Phys. Rev. E*, **71**, 036209.

Guan, S., Lai, Y.-C., Lai, C.-H., et al. 2006. Understanding synchronization induced by "common noise." *Phys. Lett. A*, **353**, 30–33.

Guckenheimer, J. and Holmes, P. 1983. *Nonlinear Oscillations, Dynamical Systems, and Bifurcations of Vector Fields*. New York: Springer-Verlag.

Güemez, J. and Matías, M. A. 1995. Modified method for synchronizing and cascading chaotic systems. *Phys. Rev. E*, **52**, R2145–R2148.

Guevara, M. R., Shrier, A., and Glass, L. 1986. Phase resetting of spontaneously beating embryonic ventricular heart cell aggregates. *Am. J. Physiol. Heart Circ. Physiol.*, **251**, H1298–H1305.

Gutiérrez, R., Amann, A., Assenza, S. et al. 2011. Emerging meso- and macroscales from synchronization of adaptive networks. *Phys. Rev. Lett.*, **107**, 234103.

Habutsu, T., Nishio, Y., Sasase, I., and Mori, S. 1991. *Advances in Cryptology – EuroCrypt'91*. Vol. 0547. Berlin: Springer-Verlag. 127–140.

Hagerstrom, A. M., Murphy, T. E., Roy, R., et al. 2012. Experimental observation of chimeras in coupled-map lattices. *Nat. Phys.*, **8**, 658–661.

Hansel, D., Mato, G., and Meunier, C. 1993. Phase dynamics for weakly coupled Hodgkin-Huxley neurons. *Europhys. Lett.*, **23**(5), 367–372.

Hansel, D., Mato, G., and Meunier, C. 1995. Synchrony in excitatory neural networks. *Neural Comput.*, **7**, 307–337.

Hayes, S., Grebogi, C., Ott, E., and Mark, A. 1993. Communicating with chaos. *Phys. Rev. Lett.*, **70**, 3031–3034.

He, R., and Vaidya, P. G. 1992. Analysis and synthesis of synchronous periodic and chaotic systems. *Phys. Rev. A*, **46**, 7387–7392.

Heagy, J. F., Platt, N., and Hammel, S. M. 1994. Characterization of on-off intermittency. *Phys. Rev. E*, **49**, 1140–1150.

Hebb, D. O. 1949. *The Organization of Behavior*. New York: Wiley.

Hernandez, R. V., Reategui, R. J., Lopez, J. H. G., and Pisarchik, A. 2008. *System and Method for Highly Safe Communication using Chaotic Signals*.

Hertz, J., Krogh, A., and Palmer, R. 1991. *Introduction to the Theory of Neural Computation*. New York: Addison-Wesley.

Hindmarsh, J. L. and Rose, R. M. 1984. A model of neuronal bursting using three coupled first order differential equations. *Proc. R. Soc. Lond. B*, **221**(1222), 87–102.

Hodgkin, A. L. and Huxley, A. F. 1952. A quantitative description of membrane current and its application to conduction and excitation in nerve. *J. Physiol.*, **117**, 500–544.

Holme, P. and Saramäki, J. 2012. Temporal networks. *Phys. Rep.*, **519**, 97–125.

Hooton, E. W. and Amann, A. 2012. Analytical limitation for time-delayed feedback control in autonomous systems. *Phys. Rev. Lett.*, **109**, 154101.

Hramov, A. E. and Koronovskii, A. A. 2005. Generalized synchronization: A modified system approach. *Phys. Rev. E*, **71**, 067201.

Hramov, A. E., Koronovskii, A. A., Kurovskaya, M. K., and Boccaletti, S. 2006. Ring intermittency in coupled chaotic oscillators at the boundary of phase synchronization. *Phys. Rev. Lett.*, **97**, 114101.

Hramov, A. E., Koronovskii, A. A., Kurovskaya, M. K., Ovchinnikov, A., and Boccaletti, S. 2007. Length distribution of laminar phases for type-I intermittency in the presence of noise. *Phys. Rev. E*, **76**(2), 026206.

Hramov, A. E., Koronovskii, A. A., Kurovskaya, M. K., and Moskalenko, O. I. 2011. Type-I intermittency with noise versus eyelet intermittency. *Phys. Lett. A*, **375**, 1646–1652.

Hu, X., Boccaletti, S., Huang, W., et al. 2014. Exact solution for first-order synchronization transition in a generalized Kuramoto model. *Sci. Rep.*, **4**, 7262.

Huerta-Cuellar, G., Pisarchik, A. N., and Barmenkov, Yu. O. 2008. Experimental characterization of hopping dynamics in a multistable fiber laser. *Phys. Rev. E*, **78**, 035202(R).

Huygens, C. 1913. *Die Pendeluhr: Horologium Oscillatorium*. Leipzig: Engelmann.

Islam, M., Islam, B., Islam, N., and Mazumder, H. P. 2013. Construction of bidirectionally coupled systems using generalized synchronization method. *Differential Geometry – Dynamical Systems*, **15**, 54–61.

Ivanchenko, M. V., Osipov, G. V., Shalfeev, V. D., and Kurths, J. 2004. Phase synchronization of chaotic intermittent oscillations. *Phys. Rev. Lett.*, **92**, 134101.

Ivanchenko, M. V., Nowotny, T., Selverston, A. I., and Rabinovich, M. I. 2008. Pacemaker and network mechanisms of rhythm generation: Cooperation and competition. *J. Theor. Biol.*, **253**, 452–461.

Izhikevich, E. M. 2007. *Dynamical Systems in Neuroscience: The Geometry of Excitability and Bursting*. Cambridge, MA: MIT Press.

Izhikevich, E. M., and Hoppensteadt, F. 2004. Classification of bursting mappings. *Int. J. Bif. Chaos*, **14**(11), 3847–3854.

Jirsa, V. K., Friedrich, R., Haken, H., and Kelso, J. A. S. 1994. A theoretical model of phase transitions in the human brain. *Biol. Cybernetics*, **71**, 27–35.

Josić, K. and Mar, D. J. 2001. Phase synchronization of chaotic systems with small phase diffusion. *Phys. Rev. E*, **64**, 056234.

Kanakidis, K., Argyris, A., and Syvridis, D. 2003. Performance characterization of high-bit-rate optical chaotic communication systems in a back to back configuration. *J. Lightwave Technol.*, **21**(3), 750–758.

Kandel, E. R., Schwartz, J. H., and Jessell, T. M. (eds). 2000. *Principles of Neural Science*. New York: McGraw-Hill.

Kaneko, K. 1986. Lyapunov analysis and information flow in coupled map lattices. *Physica D*, **23**, 436–447.

Kaneko, K. 1990. Clustering, coding, switching, hierarchical ordering, and control in network of chaotic elements. *Physica D*, **41**, 137–172.

Kaneko, K. 2015. From globally coupled maps to complex-systems biology. *Chaos*, **25**(9), 097608.

Karnatak, R., Ramaswamy, R., and Prasad, A. 2007. Amplitude death in the absence of time delays in identical coupled oscillators. *Phys. Rev. E*, **76**, 035201(R).

Katz, B. and Miledi, R. 1965. The measurement of synaptic delay, and the time course of acetylcholine release at the neuromuscular junction. *Proc. R. Soc. Lond. B*, **161**(985), 483–495.

Kazantsev, V. and Tyukin, I. 2012. Adaptive and phase selective spike timing dependent plasticity in synaptically coupled neuronal oscillators. *PLoS One*, **7**, e30411.

Kazantsev, V. B., Nekorkin, V. I., Binczak, S., Jacquir, S., and Bilbault, J. M. 2005. Spiking dynamics of interacting oscillatory neurons. *Chaos*, **15**, 23103.

Khan, M. A. and Poria, S. 2012. Generalized synchronization of bidirectionally coupled chaotic systems. *Int. J. Appl. Math. Res.*, **1**, 303–312.

Kim, D. and Kahng, B. 2007. Spectral densities of scale-free networks. *Chaos*, **17**, 026115.

Kim, H., del Genio, C. I., Bassler, K. E., and Toroczkai, Z. 2011. Constructing and sampling directed graphs with given degree sequences. *New J. Phys.*, **14**, 023012.

Kim, S., Park, S. H., and Ryu, C. S. 1997. Multistability in coupled oscillator systems with time delay. *Phys. Rev. Lett.*, **79**, 2911–2914.

Kinouchi, O. and Tragtenberg, M. H. R. 1996. Modeling neurons by simple maps. *Int. J. Bif. Chaos*, **6**, 2343–2360.

Kiss, I. Z., Kazsu, Z., and Gáspár, V. 2006. Tracking unstable steady states and periodic orbits of oscillatory and chaotic electrochemical systems using delayed feedback control. *Chaos*, **16**, 033109.

Kittel, A., Pyragas, K., and Richter, R. 1994. Prerecorded history of a system as an experimental tool to control chaos. *Phys. Rev. E*, **50**(1), 262–270.

Kitzbichler, M. G., Smith, M. L., Christensen, S. R., and Bullmore, E. 2009. Broadband criticality of human brain network synchronization. *PLoS Comput. Biol.*, **5**(3), e1000314.

Kocarev, L. and Parlitz, U. 1995. General approach for chaotic synchronization with applications to communication. *Phys. Rev. Lett.*, **74**, 5028–5031.

Kocarev, L. and Parlitz, U. 1996. Generalized synchronization, predictability, and equivalence of unidirectionally coupled dynamical systems. *Phys. Rev. Lett.*, **76**(11), 1816–1819.

Kopp, R. E., Shwom, R., Wagner, G., and Yuan, J. 2016. Tipping elements and climate-economic shocks: Pathways toward integrated assessments. *Earth's Future*, **4**, 346–372.

Koseska, A., Volkov, E., and Kurths, J. 2009. Detuning-dependent dominance of oscillation death in globally coupled synthetic genetic oscillators. *Europhys. Lett.*, **85**, 28002.

Koseska, A., Ullner, E., Volkov, E., Kurths, J., and García-Ojalvo, J. 2010a. Cooperative differentiation through clustering in multicellular populations. *J. Theor. Biol.*, **263**, 189–202.

Koseska, A., Volkov, E., and Kurths, J. 2010b. Parameter mismatches and oscillation death in coupled oscillators. *Chaos*, **20**, 023132.

Koseska, A., Volkov, E., and Kurths, J. 2013a. Oscillation quenching mechanisms: Amplitude vs. oscillation death. *Phys. Rep.*, **531**, 173–200.

Koseska, A., Volkov, E., and Kurths, J. 2013b. Transition from amplitude to oscillation death via Turing bifurcation. *Phys. Rev. Lett.*, **111**, 024103.

Krapivsky, P. L., Redner, S., and Leyvraz, F. 2000. Connectivity of growing random networks. *Phys. Rev. Lett.*, **85**, 4629–4632.

Kraut, S., Feudel, U., and Grebogi, C. 1999. On the strength of attractors in a high-dimensional system. *Phys. Rev. E*, **59**, 5253–5260.

Kruse, P. and Stadler, M. 1995. *Ambiguity in Mind and Nature: Multistable Cognitive Phenomena*. Berlin: Springer.

Kuntsevich, B. F. and Pisarchik, A. N. 2001. Synchronization effects in a dual-wavelength class-B laser with modulated losses. *Phys. Rev. E*, **64**, 046221.

Kuramoto, Y. 1975. Self-entrainment of a population of coupled non-linear oscillators. Pages 420–422 of: Araki, H. (ed.), *Lecture Notes in Physics, International Symposium on Mathematical Problems in Theoretical Physics*, vol. 39. New York: Springer-Verlag.

Kuramoto, Y. 1984. *Chemical Oscillations, Waves, and Turbulence*. Heidelberg: Springer.

Kuramoto, Y. and Battogtokh, D. 2002. Coexistence of Coherence and Incoherence in Nonlocally Coupled Phase Oscillators. *Nonl. Phenom. Compl. Syst.*, **5**, 380–385.

Kuva, S. M., Lima, G. F., Kinouchi, O., Tragtenberg, M. H. R., and Roque-da Silva, A. C. 2001. A minimal model for excitable and bursting elements. *Neurocomputing*, **38–40**, 255–261.

Kye, W. H. and Kim, C. M. 2000. Characteristic relations of type-I intermittency in the presence of noise. *Phys. Rev. E*, **62**, 6304–6307.

Laing, C. R. and Longtin, A. 2002. A two-variable model of somatic-dendritic interactions in a bursting neuron. *Bull. Math. Biol.*, **64**, 829–860.

Lang, X., Lu, Q., and Kurths, J. 2010. Phase synchronization in noise-driven bursting neurons. *Phys. Rev. E*, **82**, 021909.

Larger, L., Penkovsky, B., and Maistrenko, Y. 2013. Virtual chimera states for delayed-feedback systems. *Phys. Rev. Lett.*, **111**(5), 054103.

LeBon, G. 1896. *The Crowd: A Study of the Popular Mind*. New York: Macmillan & Co.

Lee, K. J., Kwak, Y., and Lim, T. K. 1998. Phase jumps near a phase synchronization transition in systems of two coupled chaotic oscillators. *Phys. Rev. Lett.*, **81**, 321–324.

Liu, W., Xia, J., Qian, X., and Yang, J. 2006. Antiphase synchronization in coupled chaotic oscillators. *Phys. Rev. E*, **73**, 057203.

Llinás, R. R. (ed.). 2002. *I of the Vortex: From Neurons to Self*. London: MIT Press.

Lorenz, E. N. 1963. Deterministic non-periodic flow. *Journal of the Atmospheric Sciences*, **20**, 130–141.

Luo, A. C. J. 2013. *Dynamical System Synchronization*. New York: Springer.

MacKay, R. S., and Tresser, C. 1986. Transition to topological chaos for circle maps. *Physica D*, **19**, 5–72.

Maistrenko, Y., Popovych, O., Burylko, O., and Tass, P. A. 2004. Mechanism of desynchronization in the finite-dimensional Kuramoto model. *Phys. Rev. Lett.*, **93**, 084102.

Maistrenko, Yu. L., Popovych, O. V., and Tass, P. A. 2005. Chaotic attractor in the Kuramoto model. *Int. J. Bif. Chaos*, **15**(11), 3457–3466.

Makarenko, V. and Llinás, R. 1998. Experimentally determined chaotic phase synchronization in a neuronal system. *Proc. Natl. Acad. Sci. USA*, **95**(26), 15747–15752.

Mardia, K. V. 1972. *Probability and Mathematical Statistics: Statistics of Directional Data*. London: Academic Press.

Markram, H., Lübke, J., Frotscher, M., and Sakmann, B. 1997. Regulation of synaptic efficacy by coincidence of postsynaptic APs and EPSPs. *Science*, **275**(5297), 213–215.

Martens, E. A., Thutupalli, S., Fourriere, A., and Hallatschek, O. 2013. Chimera states in mechanical oscillator networks. *Proceedings of the National Academy of Sciences*, **110**(26), 10563–10567.

Martínez-Zérega, B. E. and Pisarchik, A. N. 2012. Stochastic control of attractor preference in multistable systems. *Commun. Nonlinear Sci. Numer. Simulat.*, **17**, 4023–4028.

Matias, F. S., Carelli, P. V., Mirasso, C., and Copelli, R. M. 2011. Anticipated synchronization in a biologically plausible model of neuronal motifs. *Phys. Rev. E*, **84**, 021922.

Matias, F. S., Gollo, L. L., Carelli, P. V. Copelli, M., and Mirasso, C. R. 2013. Anticipated synchronization in neuronal motifs. *BMC Neuroscience*, **14(Suppl 1)**, P275.

Matthews, P. C. and Strogatz, S. H. 1990. Phase diagram for the collective behavior of limit-cycle oscillators. *Phys. Rev. Lett.*, **65**, 1701–1704.

Matthews, R. 1989. On the derivation of a chaotic encryption algorithm. *Cryptologia*, **XIII**, 29–42.

Maurer, J. and Libchaber, A. 1980. Effect of the Prandtl number on the onset of turbulence in liquid-He-4. *J. Phys. Lett. (Paris)*, **41**, L515–L518.

Mayol, C., Mirasso, C. R., and Toral, R. 2012. Anticipated synchronization and the predict-prevent control method in the FitzHugh–Nagumo model system. *Phys. Rev. E*, **85**, 056216.

McPhail, C. 1991. *The Myth of the Madding Crowd*. New York: Aldine de Gruyter.

Mirasso, C. R., Colet, P., and Garcia Fernandez, P. 1996. Synchronization of chaotic semi-conductor lasers: Application to encoded communications. *IEEE Photon. Tech. Lett.*, **8**, 299–301.

Mirollo, R. E. and Strogatz, S. H. 1990. Amplitude death in an array of limit-cycle oscillators. *J. Stat. Phys.*, **60**, 245–262.

Mischaikow, K., Smith, H., and Thieme, H. R. 1995. Asymptotically autonomous semi-flows: chain recurrence and Lyapunov functions. *Transactions of the American Mathematical Society*, **347**(5), 1669–1685.

Mohar, B. 1991. Eigenvalues, diameter, and mean distance in graphs. *Graph. Combinator.*, **7**, 53–64.

Monasson, R. 1999. Diffusion, localization and dispersion relations on "small-world" lattices. *Eur. Phys. J. B*, **12**, 555–567.

Motter, A. E., Zhou, C. S., and Kurths, J. 2005. Enhancing complex-network synchroniza-tion. *Eur. Phys. Lett.*, **69**, 334–340.

Moussaïd, M., Helbing, D., Garnier, S. et al. 2009. Experimental study of the behavioural mechanisms underlying self-organization in human crowds. *Proc. R. Soc. Lond. B*, **276**, 2755–2762.

Murali, K. and Lakshmanan, M. 1998. Secure communication using a compound signal from generalized synchronizable chaotic systems. *Phys. Lett. A*, **241**, 303–310.

Nayfeh, A. and Mook, D. 1979. *Nonlinear Oscillations*. New York: Wiley-Interscience.

Nelson, P. 2003. *Biological Physics: Energy, Information, Life*. New York: Freeman, W. H.

Netoff, T. I., Acker, C. D., Bettencourt, J. C., and White, J. A. 2005. Beyond two-cell net-works: experimental measurement of neuronal responses to multiple synaptic inputs. *J. Comput. Neurosci.*, **18**, 287–295.

Newhouse, S., Palis, J., and Takens, F. 1983. Bifurcations and stability of families of diffeomorphisms. *Publications Mathématiques de l'IHES*, **57**, 5–72.

Newman, M. E. J. 2003. The structure and function of complex networks. *SIAM Rev.*, **45**, 167–256.

Newman, M. E. J., Strogatz, S. H., and J., Watts D. 2001. Random graphs with arbitrary degree distributions and their applications. *Phys. Rev. E*, **64**, 026118.

Otnes, R. K. and Enochson, L. 1972. *Digital Time Series Analysis*. New York: John Wiley and Sons.

Ott, E. 2002. *Chaos in Dynamical Systems*. Second edn. New York: Cambridge University Press.

Ott, E., Grebogi, C., and Yorke, J. A. 1990. Controlling chaos. *Phys. Rev. Lett.*, **64**, 1196–1199.

Panaggio, M. J. and Abrams, D. M. 2015. Chimera states: coexistence of coherence and incoherence in networks of coupled oscillators. *Nonlinearity*, **28**(3), R67–R87.

Parlitz, U., Kocarev, L., Stojanovski, T., and Preckel, H. 1996a. Encoding messages using chaotic synchronization. *Phys. Rev. E*, **53**, 4351–4361.

Parlitz, U., Junge, L., Lauterborn, W., and Kocarev, L. 1996b. Experimental observation of phase synchronization. *Phys. Rev. E*, **54**, 2115–2117.

Pastur, L., Boccaletti, S., and Ramazza, P. L. 2004. Detecting local synchronization in coupled chaotic systems. *Phys. Rev. E*, **69**, 036201.

Pecora, L. M. and Carroll, T. L. 1990. Synchronization in chaotic systems. *Phys. Rev. Lett.*, **64**, 821–824.

Pecora, L. M. and Carroll, T. L. 1991. Driving systems with chaotic signals. *Phys. Rev. A*, **44**, 2374–2383.

Pecora, L. M. and Carroll, T. L. 1998. Master stability functions for synchronized coupled systems. *Phys. Rev. Lett.*, **80**, 2109–2112.

Pickett, J. P. et al. 2011. *American Heritage Dictionary of the English Language*. Fifth edn. Tokyo: Houghton Mifflin Company.

Pikovsky, A. and Rosenblum, M. 2008. Partially integrable dynamics of hierarchical populations of coupled oscillators. *Phys. Rev. Lett.*, **101**, 264103.

Pikovsky, A., Rosenblum, M., and Kurths, J. 2001. *Synchronization: A Universal Concept in Nonlinear Science*. New York: Cambridge University Press.

Pikovsky, A. S. 1981. A dynamical model for periodic and chaotic oscillations in the Belousov-Zhabotinsky reaction. *Phys. Lett. A*, **85**(1), 13–16.

Pikovsky, A. S., Osipov, G. V., Rosenblum, M. G., Zaks, M., and Kurths, J. 1997. Attractor-repeller collision and eyelet intermittency at the transition to phase synchronization. *Phys. Rev. Lett.*, **79**, 47–50.

Pisarchik, A. N. (ed). 2008. *Recent Achievements in Laser Dynamics: Control and Synchronization*. Kerala: Research Singpost.

Pisarchik, A. N. and Barmenkov, Yu. O. 2005. Locking of self-oscillation frequency by pump modulation in an erbium-doped fiber laser. *Opt. Commun.*, **254**, 128–137.

Pisarchik, A. N. and Corbalán, R. 1999. Parametric nonfeedback resonance in period doubling systems. *Phys. Rev. E*, **59**, 1669–1674.

Pisarchik, A. N. and Feudel, U. 2014. Control of multistability. *Phys. Rep.*, **540**, 167–218.

Pisarchik, A. N. and Jaimes-Reátegui, R. 2005a. Homoclinic orbits in a piecewise linear Rössler circuit. *J. Phys.: Conf. Ser.*, **23**, 122–127.

Pisarchik, A. N. and Jaimes-Reátegui, R. 2005b. Intermittent lag synchronization in a nonautonomous system of coupled oscillators. *Phys. Lett. A*, **338**, 141–149.

Pisarchik, A. N. and Jaimes-Reátegui, R. 2009. Control of basins of attractors in a multistable fiber laser. *Phys. Lett. A*, **374**, 228–234.

Pisarchik, A. N. and Jaimes-Reátegui, R. 2015. Deterministic coherence resonance in coupled chaotic oscillators with frequency mismatch. *Phys. Rev. E*, **92**, 050901.

Pisarchik, A. N. and Ruiz-Oliveras, F. R. 2010. Optical chaotic communication using generalized and complete synchronization. *IEEE J. Quantum Electron.*, **46**(3), 299a–299f.

Pisarchik, A. N. and Ruiz-Oliveras, F. R. 2015. Synchronization of delayed-feedback semi-conductor lasers and its application in secure communication. Pages 129–155 of: González-Aguilar, H., and Ugalde, E. (eds.), *Nonlinear Dynamics New Directions: Models and Applications*. Nonlinear Systems and Complexity, vol. 12. Switzerland: Springer.

Pisarchik, A. N., Jaimes-Reátegui, R., Villalobos-Salazar, J. R., García-Lopez, J. H., and Boccaletti, S. 2006. Synchronization of chaotic systems with coexisting attractors. *Phys. Rev. Lett.*, **96**, 244102.

Pisarchik, A. N., Jaimes-Reátegui, R., and García-Lopez, J. H. 2008a. Synchronization of coupled bistable chaotic systems: Experimental study. *Phil. Trans. Roy. Soc., Ser. A*, **366**, 459–473.

Pisarchik, A. N., Jaimes-Reátegui, R., and García-Lopez, J. H. 2008b. Synchronization of multistable systems. *Int. J. Bif. Chaos*, **18**, 1801–1819.

Pisarchik, A. N., Jaimes-Reátegui, R., Sevilla-Escoboza, R., and Boccaletti, S. 2009. Experimental approach to the study of complex networks synchronization using a single oscillator. *Phys. Rev. E*, **79**, 055202(R).

Pisarchik, A. N., Jaimes-Reátegui, R., Sevilla-Escoboza, R., Huerta-Cuillar, G., and Taki, M. 2011. Rogue waves in a multistable fiber laser. *Phys. Rev. Lett.*, **107**, 274101.

Pisarchik, A. N., Jaimes-Reátegui, R., Sevilla-Escoboza, R., and Huerta-Cuellar, G. 2012a. Multistate intermittency and extreme pulses in a fiber laser. *Phys. Rev. E*, **86**, 056219.

Pisarchik, A. N., Jaimes-Reátegui, R., Sevilla-Escoboza, J. R., Ruiz-Oliveras, F. R., and García-López, J. H. 2012b. Two-channel opto-electronic chaotic communication system. *Journal of Franklin Institut*, **349**, 3194–3202.

Pisarchik, A. N., Bashkirtseva, I., and Ryashko, L. 2017. Chaos can imply periodicity in coupled oscillators. *Europhys. Lett.*, **117**, 40005.

Platt, E. A., Spiegel, E. A., and Tresser, C. 1993. On-off intermittency: A mechanism for bursting. *Phys. Rev. Lett.*, **70**(3), 279–282.

Pomeau, Y., and Manneville, P. 1980. Intermittent transition to turbulence in dissipative dynamical systems. *Commun. Math. Phys.*, **74**(2), 189–197.

Popovych, O. V., Maistrenko, Y. L., and Tass, P. A. 2005. Phase chaos in coupled oscillators. *Phys. Rev. E*, **71**, 065201.

Poria, S. 2007. The linear generalized chaos synchronization and predictability. *Int. J. of Appl. Mech. Eng.*, **12**, 879–885.

Postnov, D. E., Vadivasova, T. E., Sosnovtseva, O. V., Balanov, A. G., and Anishchenko, V. S. 1999. Role of multistability in the transition to chaotic phase synchronization. *Chaos*, **9**, 227–232.

Press, W. H., Teukolsky, S. A., Vetterling, W. T., and Flannery, B. P. 2007. *Numerical Recipes: The Art of Scientific Computing*. Third edn. Cambridge: Cambridge University Press.

Prigogine, I. and Stengers, I. 1965. *Modulation, Noise, and Spectral Analysis*. New York: McGraw-Hill.

Pyragas, K. 1992. Continuous control of chaos by self-controlling feedback. *Phys. Lett. A*, **170**, 421–428.

Pyragas, K. 1993. Predictable chaos in slightly perturbed unpredictable chaotic systems. *Phys. Lett. A*, **181**, 203–210.

Pyragas, K. 1996. Weak and strong synchronization of chaos. *Phys. Rev. E*, **54**(5), R4508–R4511.

Pyragas, K. 1997. Conditional Lyapunov exponents from time series. *Phys. Rev. E*, **56**, 5183–5186.

Pyragas, K. and Novičenko, V. 2013. Time-delayed feedback control design beyond the odd-number limitation. *Phy. Rev. E*, **88**, 012903.

Pyragas, K. and Tamaševičius, A. 1993. Experimental control of chaos by delayed self-controlling feedback. *Phy. Lett. A*, **180**(1), 99–102.

Pyragiené, T. and Pyragas, K. 2013. Anticipating spike synchronization in nonidentical chaotic neurons. *Nonlinear Dynamics*, **74**, 297–306.

Quian Quiroga, R., Arnhold, J., and Grassberger, P. 2000. Learning driver-response relationships from synchronization patterns. *Phys. Rev. E*, **61**(5), 5142–5148.

Quian Quiroga, R., Kraskov, A., Kreuz, T., and Grassberger, P. 2002. Performance of different synchronization measures in real data: A case study on electroencephalographic signals. *Phys. Rev. E*, **65**(5), 041903.

Rabinovich, M. I. and Abarbanel, H. D. I. 1999. The role of chaos in neural systems. *Neuroscience*, **87**(1), 5–14.

Rabinovich, M. I. and Trubetskov, D. I. 1989. *Introduction to the Theory of Oscillations and Waves*. Amsterdam: Kluwer.

Ramana Reddy, D. V., Sen, A., and Johnston, G. L. 1998. Time delay induced death in coupled limit cycle oscillators. *Phys. Rev. Lett.*, **80**, 5109–5112.

Rex, C. S., Colgin, L. L., Jia, Y. et al. 2009. Origins of an intrinsic hippocampal EEG pattern. *PLoS One*, **4**(11), e7761.

Rodrigues, F. A., Peron, T. K. D., Ji, P., and Kurths, J. 2016. The Kuramoto model in complex networks. *Phys. Rep.*, **610**, 1–98.

Rosa, E., Ott, E., and Hess, M. H. 1998. Transition to phase synchronization of chaos. *Phys. Rev. Lett.*, **80**, 1642–1645.

Rosenblum, M. G., Pikovsky, A. S., and Kurths, J. 1996. Phase synchronization of chaotic oscillators. *Phys. Rev. Lett.*, **76**, 1804–1897.

Rosenblum, M. G., Pikovsky, A. S., and Kurths, J. 1997. From phase to lag synchronization in coupled chaotic oscillators. *Phys. Rev. Lett.*, **78**, 4193–4196.

Rössler, O. E. 1976. An equation for continuous chaos. *Phys. Lett. A*, **57**, 397–398.

Rössler, O. E. 1977. Chaos in abstract kinetics 2 prototypes. *Bull. Math. Biol.*, **39**, 275–289.

Roy, P. K., Chakraborty, S., and Dana, S. K. 2003. Experimental observation on the effect of coupling on different synchronization phenomena in coupled nonidentical Chua's oscillators. *Chaos*, **13**, 342–355.

Ruiz-Oliveras, F. R. and Pisarchik, A. N. 2006. Phase-locking phenomenon in a semiconductor laser with external cavities. *Opt. Express*, **14**, 12859–12867.

Ruiz-Oliveras, F. R. and Pisarchik, A. N. 2009. Synchronization of semiconductor lasers with coexisting attractors. *Phys. Rev. E*, **79**, 016202.

Rulkov, N. F. 2001. Regularization of synchronized chaotic bursts. *Phys. Rev. Lett.*, **86**, 183–186.

Rulkov, N. F. 2002. Modeling of spiking-bursting neural behavior using two-dimensional map. *Phys. Rev. E*, **65**, 041922.

Rulkov, N. F., Sushchik, M. M., Tsimring, L. S., and Abarbanel, H. D. I. 1995. Generalized synchronization of chaos in directionally coupled chaotic systems. *Phys. Rev. E*, **51**, 980–994.

Rulkov, N. F., Timofeev, I., and Bazhenov, M. 2004. Oscillations in large-scale cortical networks: map-based model. *J. Comput. Neurosci.*, **17**, 203–223.

Sakaguchi, H. and Kuramoto, Y. 1986. A soluble active rotater model showing phase transitions via mutual entertainment. *Progr. Theor. Phys.*, **76**(3), 576–581.

Sausedo-Solorio, J. M. and Pisarchik, A. N. 2011. Dynamics of unidirectionally coupled Hénon maps. *Phys. Lett. A*, **375**, 3677–3681.

Saxena, G., Prasad, A., and Ramaswamy, R. 2012. Amplitude death: The emergence of stationarity in coupled nonlinear systems. *Phys. Rep.*, **521**, 205–230.

Schäfer, C., Rosenblum, M. G., Kurths, J., and Abel, H. H. 1998. Heartbeat synchronized with ventilation. *Nature*, **392**, 239–240.

Schikora, S., Hövel, P., Wünsche, H.-J., Schöll, E., and Henneberger, F. 2006. All-optical noninvasive control of unstable steady states in a semiconductor laser. *Phys. Rev. Lett.*, **97**, 213902.

Schöll, E. and Schuster, H. G. 2008. *Handbook of Chaos Control*. Second edn. City: Wiley-VCH.

Schuster, H. G. and Just, W. 2005. *Deterministic Chaos: An Introduction*. Fourth edn. City: Wiley-VCH.

Sevilla-Escoboza, R., Buldú, J. M., Pisarchik, A. N., Boccaletti, S., and Gutiérrez, R. 2015. Synchronization of intermittent behavior in ensembles of multistable dynamical systems. *Phys. Rev. E*, **91**, 039202.

Seydel, R. 1988. *From Equilibrium to Chaos: Practical Bifurcation and Stability Analysis.* New York: Elsevier.

Shannon, C. E. 1949. Communication theory of secrecy systems. *Bell Syst. Tech. J.*, **28**, 656–715.

Shen-Orr, S. S., Milo, R., Mangan, S., and Alon, U. 2002. Network motifs in the transcriptional regulation network of Escherichia coli. *Nat. Genet.*, **31**, 64–68.

Shilnikov, A. and Rulkov, N. F. 2004. Subthreshold oscillations in a map-based neuron model. *Phys. Lett. A*, **328**, 177–184.

Shilnikov, A., Gordon, R., and Belykh, I. 2008. Polyrhythmic synchronization in bursting networking motifs. *Chaos*, **18**, 037120.

Shilnikov, A. L. and Rulkov, N. F. 2003. Origin of chaos in a two-dimensional map modeling spiking-bursting neural activity. *Int. J. Bif. Chaos*, **13**, 3325–3340.

Shilnikov, L. P. 1965. A case of the existence of a countable number of periodic motions. *Sov. Math. Doklady*, **6**, 163–166.

Shima, S. and Kuramoto, Y. 2004. Rotating spiral waves with phase-randomized core in nonlocally coupled oscillators. *Phys. Rev. E*, **69**, 036213.

Shore, K. A., Spencer, P. S., and Pierce, I. 2008. Synchronization of chaotic semiconductor lasers. Pages 79–104 of: Pisarchik, A. N. (ed.), *Recent Advances in Laser Dynamics: Control and Synchronization*. Kerala: Research Singpost.

Sieber, J., Gonzalez-Buelga, A., Neild, S. A., Wagg, D. J., and Krauskopf, B. 2008. Experimental continuation of periodic orbits through a fold. *Phys. Rev. Lett.*, **100**, 244101.

Simonov, A. Yu., Gordleeva, S. Yu., Pisarchik, A. N., and Kazantsev, V. B. 2013. Synchronization with an arbitrary phase shift in a pair of synaptically coupled neural oscillators. *J. Exper. Theor. Phys. Lett.*, **98**(10), 632–637.

Strogatz, S. H. 2003. *Sync: The Emerging Science of Spontaneous Order*. New York: Hyperion Press.

Strogatz, S. H. 2015. *Nonlinear Dynamics and Chaos: With Applications to Physics, Biology, Chemistry, and Engineering*. Second edn. Boulder, CO: Westview Press.

Strogatz, S. H., and Mirollo, R. E. 1991. Stability of incoherence in a population of coupled oscillators. *J. Stat. Phys.*, **63**, 613–635.

Sun, X., Lu, Q., Kurths, J., and Wang, Q. 2009. Spatiotemporal coherence resonance in a map lattice. *Int. J. Bif. Chaos*, **19**(2), 737–743.

Takens, F. 1981. Detecting strange attractors in turbulence. Pages 366–381 of: Rand, D. A. and Yong, L.-S. (eds.), *Lecture Notes in Mathematics*, vol. 898. New York: Springer-Verlag.

Tass, P., Rosenblum, M. G., Weule, J., Kurths, J. et al. 1998. Detection of n:m phase locking from noisy data: application to magnetoencephalography. *Phys. Rev. Lett.*, **81**, 3291–3294.

Terry, J. R. and VanWiggeren, G. D. 2001. Chaotic communication using generalized synchronization. *Chaos, Solitons and Fractals*, **12**, 145–152.

Thompson, J. M. T. and Stewart, H. B. 1986. *Nonlinear Dynamics and Chaos*. Chichester: Wiley.

Tinsley, M. R., Nkomo, S., and Showalter, K. 2012. Chimera and phase-cluster states in populations of coupled chemical oscillators. *Nat. Phys.*, **8**(9), 662–665.

Turing, A. M. 1952. The chemical basis of morphogenesis. *Philos. Trans. R. Soc. Lond. B*, **237**, 37–72.

Turner, R. H. and Killia, L. M. 1993. *Collective Behavior*. Englewood Cliffs, NJ: Prentice Hall.

Uhm, W., Astakhov, V., Akopov, A., and Kim, S. 2003. Multistability formation and loss of chaos synchronization in coupled period-doubling systems. *Int. J. Mod. Phys. B*, **17**, 4013–4022.

Valverde, S. and Solé, R. V. 2005. Network motifs in computational graphs: A case study in software architecture. *Phys. Rev. E*, **72**, 026107.

Van der Pol, B. and Van der Mark, J. 1927. Frequency demultiplication. *Nature*, **120**, 363–364.

Verhults, F. 1990. *Nonlinear Differential Equations and Dynamical Systems*. Berlin: Springer.

Voss, H. U. 2000. Anticipating chaotic synchronization. *Phys. Rev. E*, **61**, 5115–5119.

Voss, H. U. 2001. Dynamic long-term anticipation of chaotic states. *Phys. Rev. Lett.*, **87**, 014102.

de Vries, G. 2001. Bursting as an emergent phenomenon in coupled chaotic maps. *Phys. Rev. E*, **64**, 051914.

Wallenstein, G. V., Scott Kelso, J. A., and Bressler, S. L. 1995. Phase transitions in spatiotemporal patterns of brain activity and behavior. *Physica D*, **84**, 626–634.

Wang, H., Lu, Q. and Wang, Q. 2008. Bursting and synchronization transition in the coupled modified ML neurons. *Commun. Nonlinear Sci. Numer. Simulat.*, **13**(8), 1668–1675.

Warren, R. M. (ed). 1999. *Auditory Perception: A New Analysis and Synthesis*. Cambridge: Cambridge University Press.

Watts, D. J. and Strogatz, S. H. 1998. Collective dynamics of "small-world" networks. *Nature*, **393**, 440–442.

Willems, J. L. 1970. *Stability Theory of Dynamical Systems*. New York: Wiley.

Winfree, A. T. 1984. The prehistory of the Belousov–Zhabotinsky oscillator. *J. Chem. Educ*, **61**(8), 661–663.

Wolfrum, M. and Omel'chenko, O. E. 2011. Chimera states are chaotic transients. *Phys. Rev. E*, **84**, 015201.

Yamaguchi, Y. and Shimizu, H. 1984. Theory of self-synchronization in the presence of native frequency distribution and external noises. *Physica D*, **11**(1), 212–226.

Yamasue, K., Kobayashi, K., Yamada, H., Matsushige, K., and Hikihara, T. 2009. Controlling chaos in dynamic-mode atomic force microscope. *Phys. Lett. A*, **373**, 3140–3144.

Yanchuk, S. and Kapitaniak, T. 2003. Chaos-hyperchaos transition in coupled Rössler systems. *Phys. Lett. A*, **290**, 139–144.

Yeldesbay, A., Pikovsky, A., and Rosenblum, M. 2014. Chimeralike states in an ensemble of globally coupled oscillators. *Phys. Rev. Lett.*, **112**, 144103.

Yu, L., Ott, E., and Chen, Q. 1990. Transition to chaos for random dynamical systems. *Phys. Rev. Lett.*, **65**, 2935–2938.

Zakharova, A., Kapeller, M., and Schöll, E. 2014. Chimera death: Symmetry breaking in dynamical networks. *Phys. Rev. Lett.*, **112**, 154101.

Zhan, M., Hu, G., He, D. H., and Ma, W. Q. 2001. Phase locking in on-off intermittency. *Phys. Rev. E*, **64**, 066203.

Zou, W., Senthilkumar, D. V., Koseska, A., and Kurths, J. 2013. Generalizing the transition from amplitude to oscillation death in coupled oscillators. *Phys. Rev. E*, **88**, 050901(R).

Index